Insecticides and Pesticides: Methods for Crop Protection

Insecticides and Pesticides: Methods for Crop Protection

Zoe Wordsworth

R CALLISTO REFERENCE

www.callistoreference.com

Callisto Reference,
118-35 Queens Blvd., Suite 400,
Forest Hills, NY 11375, USA

Visit us on the World Wide Web at:
www.callistoreference.com

ISBN: 978-1-64116-536-5 (Hardback)

Cataloging-in-Publication Data

Insecticides and pesticides : methods for crop protection / Zoe Wordsworth.
    p. cm.
Includes bibliographical references and index.
ISBN 978-1-64116-536-5
1. Insecticides. 2. Pesticides. 3. Plants, Protection of. 4. Crops.
5. Agriculture. I. Wordsworth, Zoe.
SB951.5 .I57 2022
632.951 7--dc23

Table of Contents

Preface

This book has been written, keeping in view that students want more practical information. Thus, my aim has been to make it as comprehensive as possible for the readers. I would like to extend my thanks to my family and co-workers for their knowledge, support and encouragement all along.

Insecticides and pesticides are substances that are used to kill insects and control pests respectively. Insecticides involve the usage of ovicides against the eggs of the insects and larvicides against insect larvae. They are used in various sectors such as agriculture, medicine, industries, etc. They play an important role in increasing agricultural productivity. Insecticides are classified into two major groups, namely, systemic and contact insecticides. Systemic insecticides have residual and long term activity and contact insecticides are lethal or harmful to the insects when they come in direct contact with it. Pesticides are chemical and biological agents that block and kill pests. Some of the common types of pesticides are herbicides, insecticides, nematicides, molluscicides, piscicides and insect repellents. They also include rodenticides, bactericides and animal repellents. The topics included in this book on insecticides and pesticides are of utmost significance and bound to provide incredible insights to readers. It is compiled in such a manner, that it will provide in-depth knowledge in this area. This book is an essential guide for both academicians and those who wish to pursue this discipline further.

A brief description of the chapters is provided below for further understanding:

Chapter – What are Pesticides?

Pesticides are the substances that are used to control pests, including insects, rodents, fungi and weed. Acaricide, chlorpyrifos and pyriproxyfen are some of the pesticides which help in producing more crops with less land, ensuring better harvests and help in conserving the environment. The topics elaborated in this chapter will help in gaining a better perspective about pesticides.

Chapter – Types of Pesticides

Herbicides, biopesticides and organic pesticides are the three main types of pesticides. Herbicides are the compounds that are used to control weeds and biopesticides are the pesticides derived from naturally occuring materials such as plants, animals and certain minerals. This chapter has been carefully written to provide an easy understanding of these various types of pesticides as well as their applications.

Chapter – Insecticides and its Types

The agents of chemical and biological origin which control or kill insects are referred to as insecticides. Some of the common types of insecticides are organochlorines, organophosphates, organosulfur formamidines, dinitrophenols, fiproles, etc. All these types of insecticides have been carefully analyzed in this chapter.

Chapter – Organochlorine Pesticides and Insecticides

Organochlorine pesticides and insecticides are chlorinated hydrocarbons which are widely used all over the world. It includes DDT, methoxychlor, dieldrin, chlordane, toxaphene, mirex, kepone, lindane, and benzene hexachloride. The topics elaborated in this chapter will help in gaining a better perspective about organochlorine pesticides and insecticides.

Chapter – Health and Environmental Effects

Pesticides and insecticides have a widespread impact on the environment as well as human health. They have devastating effects on soil, water, air and wildlife and cause various skin and eye diseases in humans. The chapter closely examines these environmental and health impacts of pesticides and insecticides to gain a better understanding of the subject.

Zoe Wordsworth

What are Pesticides?

Pesticides are the substances that are used to control pests, including insects, rodents, fungi and weed. Acaricide, chlorpyrifos and pyriproxyfen are some of the pesticides which help in producing more crops with less land, ensuring better harvests and help in conserving the environment. The topics elaborated in this chapter will help in gaining a better perspective about pesticides.

Pesticides are chemical substances that are meant to kill pests. In general, a pesticide is a chemical or a biological agent such as a virus, bacterium, antimicrobial, or disinfectant that deters, incapacitates, kills, pests.

This use of pesticides is so common that the term pesticide is often treated as synonymous with plant protection product. It is commonly used to eliminate or control a variety of agricultural pests that can damage crops and livestock and reduce farm productivity. The most commonly applied pesticides are insecticides to kill insects, herbicides to kill weeds, rodenticides to kill rodents, and fungicides to control fungi, mold, and mildew.

Pesticides are not recent inventions. Many ancient civilizations used pesticides to protect their crops from insects and pests. Ancient Sumerians used elemental sulfur to protect their crops from insects. Whereas, Medieval farmers experimented with chemicals using arsenic, lead on common crops.

The Chinese used arsenic and mercury compounds to control body lice and other pests. While, the Greeks and Romans used oil, ash, sulfur, and other materials to protect themselves, their livestock, and their crops from various pests.

Meanwhile, in the nineteenth century, researchers focused more on natural techniques involving compounds made with the roots of tropical vegetables and chrysanthemums. In 1939, Dichloro-Diphenyl-Trichloroethane (DDT) was discovered, which has become extremely effective and rapidly used as the insecticide in the world. However, twenty years later, due to biological effects and human safety, DDT has been banned in almost 86 countries.

Types of Pesticides

These are grouped according to the types of pests which they kill.

Grouped by types of pests they kill:

- Insecticides – Insects.

- Herbicides – Plants.

- Rodenticides – Rodents (rats & mice).

- Bactericides – Bacteria.

- Fungicides – Fungi.

- Larvicides – Larvae.

Based on how biodegradable they are:

Pesticides can also be considered as:

- Biodegradable: The biodegradable kind is those which can be broken down by microbes and other living beings into harmless compounds.

- Persistent: While the persistent ones are those which may take months or years to break down.

Another way to classify these is to consider those that are chemical forms or are derived from a common source or production method.

Chemically-related pesticides:

- Organophosphate: Most organophosphates are insecticides; they affect the nervous system by disrupting the enzyme that regulates a neurotransmitter.

- Carbamate: Similar to the organophosphorus pesticides, the carbamate pesticides also affect the nervous system by disrupting an enzyme that regulates the neurotransmitter. However, the enzyme effects are usually reversible.

- Organochlorine insecticides: They were commonly used earlier, but now many countries have been removed Organochlorine insecticides from their market due to their health and environmental effects and their persistence (e.g., DDT, chlordane, and toxaphene).

- Pyrethroid: These are a synthetic version of pyrethrin, a naturally occurring pesticide, found in chrysanthemums (Flower). They were developed in such a way as to maximise their stability in the environment.

- Sulfonylurea herbicides: The sulfonylureas herbicides have been commercialized for weed control such as pyrithiobac-sodium, cyclosulfamuron, bispyribac-sodium, terbacil, sulfometuron-methyl Sulfosulfuron, rimsulfuron, pyrazosulfuron-ethyl, imazosulfuron, nicosulfuron, oxasulfuron, nicosulfuron, flazasulfuron, primisulfuron-methyl, halosulfuron-methyl, flupyrsulfuron-methyl-sodium, ethoxysulfuron, chlorimuron-ethyl, bensulfuron-methyl, azimsulfuron, and amidosulfuron.

- Biopesticides: The biopesticides are certain types of pesticides derived from such natural materials as animals, plants, bacteria, and certain minerals.

Examples of Pesticides

Examples of pesticides are fungicides, herbicides, and insecticides. Examples of specific synthetic chemical pesticides are glyphosate, Acephate, Deet, Propoxur, Metaldehyde, Boric Acid, Diazinon, Dursban, DDT, Malathion, etc.

Benefits of Pesticides

The major advantage of pesticides is that they can save farmers. By protecting crops from insects and other pests. However, below are some other primary benefits of it.

- Controlling pests and plant disease vectors.

- Controlling human/livestock disease vectors and nuisance organisms.

- Controlling organisms that harm other human activities and structures.

Effects of Pesticides

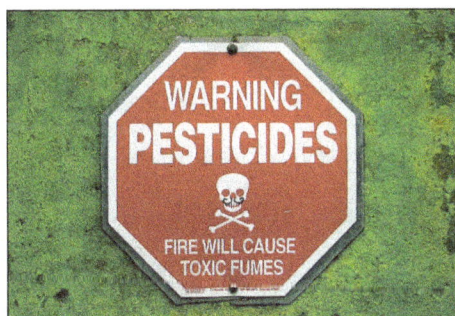

- The toxic chemicals in these are designed to deliberately release into the environment. Though each pesticide is meant to kill a certain pest, a very large percentage of pesticides reach a destination other than their target. Instead, they enter the air, water, sediments, and even end up in our food.

- Pesticides have been linked with human health hazards, from short-term impacts such as headaches and nausea to chronic impacts like cancer, reproductive harm.

- The use of these also decreases the general biodiversity in the soil. If there are no chemicals in the soil there is a higher soil quality, and this allows for higher water retention, which is necessary for plants to grow.

PESTICIDE APPLICATION

Pesticide application plays an important role in pest management. Proper technique of application of pesticide and the equipment used for applying pesticide are vital to the success of pest control operations. The application of pesticide is not merely the operation of sprayer or duster. It has to be coupled with a thorough knowledge of the pest problem. The use of pesticides involves knowledge not only of application equipment, but of pest management as well.

The main purpose of pesticide application technique is to cover the target with maximum efficiency and minimum efforts to keep the pest under control as well as minimum contamination of non-targets. All pesticides are poisonous substances and they can cause harm to all living things. Therefore their use must be very judicious. The application techniques ideally should be target oriented so that safety to the non-targets and the environment is ensured. Therefore, proper selection of application equipment, knowledge of pest behavior and skillful dispersal methods are vital. The complete knowledge of pest problem is important to define the target i.e., location of the pest (on foliage, under the leaves, at root zone etc.). The most susceptible stage of the pest for control measures will help to decide the time of application. The requirement of coverage and spray droplet size depends upon the mobility and size of the pest. The mode of action of pesticide, its relative toxicity and other physicochemical properties, help to decide the handling precautions, agitation requirement etc. Further the complete knowledge of the equipment is necessary to develop desired skill of operation, to select and to estimate the number and type of equipment needed to treat the crop in minimum time and to optimize use of the equipment.

Thus the following aspects must be considered for a successful pest control programme:

- Knowledge of pest problem:
  - Where is the pest location? - To define the target.
  - What is the most susceptible stage for control? - To decide the time of application.
  - What is the mobility of the pest? - To define the coverage requirements and droplet size.
- Knowledge of pesticides:
  - What is the mode of action? - To define the application technique.
  - What is the phytotoxicity? - To define the calibration requirement.
  - What is mammalian toxicity? - To take up necessary precautions in handling.
- Knowledge of formulations:
  - What is the solubility? - To define the agitation requirements.
  - How should it be mixed with water? - To collect suitable measure and water buckets and tools etc.
- Knowledge of techniques and equipments:
  - How should it be operated and maintained? - To operate the equipment without field troubles.
  - What are the capabilities? - To estimate sufficient number of equipment needed.
  - What adjustments are necessary? - To get an optimum use of the equipment.
  - What technique is to be adopted? - To select suitable equipment.

Objective of Pesticide Application

The objective of the application of pesticide is to keep the pest under check. The pest population has to be kept suppressed to minimum biological activities to avoid economic loss of crop yields. Thorough killing of pest or eradication of pest is neither practical nor necessary. The objective of pesticide application besides keeping the pest population under check should also be to avoid pollution and damage to the non-targets.

The success of pest control operations by pesticide application greatly depends on the following factors:

- Quality of pesticide,
- Timing of application,
- Quality of application and coverage.

Different types of pesticides are used for controlling various pests. For example Insecticides are applied against insect pests, Fungicides against crop diseases, Herbicides against weeds etc. In order to protect the crop losses. But it is essential that besides choosing an appropriate pesticide for application it has to be a quality product i.e., proper quantity of pesticide active ingredient (a.i) must be ensure that the quantity is maintained in production and marketing of pesticide formulations.

The application of pesticide is very successful when applied at the most susceptible stage of the pest. If the timing of pesticide application is carefully considered and followed, the results will be good pest control and economy. Therefore for large area treatment careful selection of equipment becomes necessary so that within the available 'Time' the area could be treated.

Even though good quality pesticide is used and optimum timing for the application of pesticide is also adopted; unless the pesticide is applied properly it will not yield good results. Therefore, the quality of application of pesticides is very important in pest control operations. Adherence to the following points can ensure it:

- Proper dosage should be applied evenly.
- The toxicant should reach the target.
- Proper droplet size.
- Proper density of droplet on the target.

The dosage recommendation are generally indicated for acre or hectare e.g. kg/ha or lit/ha or gm ai/ha. It should be properly understood and the exact quantities of the formulated pesticide should be applied.

Pesticides are dispersed by different methods like spraying, dusting etc. For spraying of pesticides different types of nozzles such as hydraulic, air blast, centrifugal and heat energy type are used. Water is a common carrier of pesticides but air or oils are also used as carriers. Selection of proper droplet is an important consideration. The shape, size and surface of the target vary greatly. For spraying against flying insects, the hydraulic nozzles will not be effective. Here we need fine size spray particles to remain airborne for longer time. However, for weed control operation usually the requirement is drift free application or coarse spray droplets. Adequate number of spray droplets should be deposited necessarily. For fungicide application the number of droplets deposited per unit area should be more and may be for translocated herbicide application it can be less in number. It may need fewer numbers of droplets to be deposited in case of highly mobile (crawling) insect pest.

The pesticides are formulated in liquid form, dust powder or granule forms such that it makes possible to apply small quantities of pesticides over large area. Some of the pesticides are applied as low as few gram a.i. per hectare. Therefore adoption of proper Application Technique is vital for uniform depositing of pesticide. The method of setting the pesticide application equipment to ensure even distribution of certain quantity of pesticide over the desired area is called Calibration.

Spraying Techniques

Most of the pesticides are applied as sprays:

The liquid formulations of pesticide either diluted (with water, oil) or directly are applied in small drops to the crop by different types of sprayers. Usually the EC formulations, wettable powder formulations are diluted suitably with water which is a common carrier of pesticides. In some cases however, oil is used as diluent or carrier of pesticides.

The important factors for spray volume consideration are:

The volume of spray liquid required for certain area depends upon the spray type and coverage, total target area, size of spray droplet and number of spray droplets. It is obvious that if the spray droplets are coarse-size then the spray volume required will be larger than the small size spray droplets. Also if the thorough coverage (e.g. both the sides of leaves) is necessary then the spray volume requirement has to be more.

On the basis of volume of spray-mix the technique of spraying is classified as:

- High volume spraying.

- Low volume spraying.

- Ultra low volume spraying.

The range of volume of spray mix in each of the above case is arbitrary. Usually for field crop spraying the following spray volume ranges are taken as guide.

High Volume Spraying	300 - 500 L/ha
Low Volume Spraying	50 - 150 L/ha
Ultra Low Volume Spraying	< 5 L/ha

There is distinct advantage in the case of lower volume of application over the high volume application. The higher the volume to be applied the more the time, the more the labor and the more the cost of application due to labor cost. However the lower volume applications are concentrated spraying of pesticide which should also be considered properly.

Sprayers (Hydraulic Energy)	
Manually operated	Powered operated
1. Syringes, slide pump	1. High pressure sprayer (hand carried type)
2. Stirrup pumps	2. High pressure trolley/Barrow mounted
3. Knap sack or shoulder-slung: • Lever operated K.S. sprayer • Piston pump type • Diaphragm pump type	3. Tractor mounted/trailed sprayer
4. Compression sprayer: • Hand compression sprayer • Conventional type • Pressure retaining type	4. High pressure knap sack sprayer
5. Stationary type: • Foot operated sprayer • Rocker sprayer	5. Air craft, aerial spraying (Fixed wing, helicopter)

Sprayers (Gaseous energy)	
Manually operated	Powered operated
1. Hand held type	1. Knap sack, motorized type 2. Hand/Stretcher carried type 3. Tractor mounted

Classification of Plant Protection Equipments

Sprayers (Centrifugal Energy):

1. Hand held battery operated ULV sprayer.

2. Knapsack motorized type.

3. Tractor/vehicle mounted ULV sprayer.

4. Aircraft ULV sprayer.

Other Sprayers:

1. Aerosol sprayers.

2. Liquefied-gas type dispensers.

3. Fogging machines.

4. Exhaust Nozzle Sprayer.

Dusting Equipment	
Manually operated	Powered operated
1. Plunger duster	1. Knapsack motorized duster
2. Bellow duster	2. High pressure trolley/Barrow mounted

3. Rotary duster:
 	
  • Belly mounted model
 	
  • Shoulder-slung model
	3. Tractor mounted/trailed duster

	4. Aircraft

Granule Applicator	
1. Broad-casting tins	1. Knapsack motorized type
2. Knapsack Rotary granule	2. Tractor mounted/trailed duster
	3. Aircraft

Standardization and Testing Methods of Plant Protection Equipment

The object of proper pesticide application cannot be achieved without good quality Plant Protection Equipment. A well designed machine shall be efficient as far as pesticide distribution and delivery to the target in minimum time with minimum wastages is concerned. Therefore, the machines should be tested to ascertain that they are:

- Efficient,
- Reliable,
- Long lasting,
- Comfortable to operate,
- Minimum field problems.

The machines should meet certain minimum requirements of performance, efficiency and reliability. For this it is essential that standard specifications are laid down so that it will have the above said qualities. The equipment standard specification parameters are:

- Material of construction,
- Dimensions,
- Ergonomics,
- Stability,
- Safety,
- Interchangeability,
- Performance,

- Strength, Reliability,

- Workmanship, Finishing,

Besides the equipment, the components of the system should be standardized and tested such as:

- Nozzles, Cut-off devices, Lances etc.

Some important aspects of the specifications and testing methods are as under:

- Compression sprayer: The routine specifications of material of construction, dimensions, workmanship are included. As the tank of the sprayers is subjected to high pressure, a tank fatigue test is recommended. The spray tank is pressurized by hydraulic force and depressurized. Such 1200 cycles of pressurization are imposed on the tank during which it should not leak. Similarly the impact strength of the sprayer is tested by dropping the filled and pressurized sprayer from a given height in different positions. Also the straps are tested for supporting the weight of the sprayer when it falls from a certain height.

- Knapsack sprayer: Besides the material of construction, dimensions, capacity, other aspects of performance are specified. The volumetric efficiency of the pump should be above 80%, the ratio between the pump volume per stroke and the pressure chamber volume should be minimum 1:8. The operating lever movement for full pump stroke should not be more than 350 for each movement i.e. upward and downward. The pump discharge rate at 16±1 stroke per minute at 40 psi pressure should be minimum 500 ml/min. The reliability test for 48 hrs continuous working of the sprayer is also recommended.

- Foot sprayer & Rocking sprayer: The specifications in respect of material of construction, dimension, workmanship and finishing are standardized. The volumetric efficiency of the pump should be minimum 80%. Other parameters like discharge rate test, ratio of volume and pressure chamber volume, leakage test etc., are considered. The reliability test of 48 hrs of continuous working of the sprayer is recommended.

- Motorized sprayer: The spraying systems except the engines are covered in the specifications. The usual specification of material, dimension, capacity, discharge rate is standardized. The air delivery volume and velocity of air at nozzle are also specified. The reliability test and fuel consumption test are recommended.

- Hydraulic nozzle & Spray lance: The spray discharge rate and other physical parameters viz. spray angle and spray distribution pattern are specified. The nozzle tip abrasion test is also recommended to ascertain the reliability of performance of hydraulic nozzles.

- Cut-off device: The reliability test of cut-off device for 5000 cycles of operation spraying with fine silica powder (abrasive) is recommended. The test for measurement of effort to actuate the lever of the trigger is also specified.

For the above described testing of the Plant Protection Equipment and components the following test rigs are used:

- Tank fatigue strength test-rig.

- Knapsack sprayer test-rig.

- Spray pattern test-rig.

- Nozzle abrasion test-rig.

- Cut-off device test-rig.

- Impact strength test-rig.

Problems of Maintenance and Repairs of Plant Protection Equipment

Plant protection machines in general are not well maintained regularly either in godowns/depots where they are stored or in the field where they are used. Life of a machine depends entirely on its care and maintenance. Even though machines are made with high standards of skill and workmanship, they can easily be ruined due to improper care and maintenance. Good and constant performance from machines can be obtained only when they are used and serviced periodically. The purpose of maintaining a machine is for increasing the useful life of the machine and to be available in working order whenever put to use. The maintenance of a machine involves proper care, operation, servicing, repair and keeping it in good working order.

Maintenance

Normal maintenance jobs include cleaning the equipment and applying necessary lubricating oils and greases to the rubbing and moving parts. If this normal maintenance is neglected the machine gets rusted and moving parts wear out quickly resulting in loss of efficiency, frequent replacement of spare parts and finally uneconomical working.

Besides the normal maintenance as above, special care has to be taken for maintaining the plant protection equipment. The pesticide formulations are chemically aggressive on metals, etc. The cleaning and washing of the chemical tank, discharge lines, nozzles, etc., are to be done regularly after the day's spraying work is completed otherwise the residues of chemicals used for spraying acts on the parts and causes corrosion and deterioration of materials.

If this aspect of thorough cleaning is not done on the plant protection machine, even though it is made of with high standard materials, it will not serve its normal life and would lead to premature condemnation.

Maintenance Job for Hand Operated Equipment

- Cleaning the chemical tanks, hoses, valves and nozzles etc. and flushing sufficiently to avoid pesticide residue which is corrosive.

- Cleaning the machine equally well from outside also as it is contaminated due to leakage, spilling of pesticide.

- Lubricating suitably the pump parts like piston, cylinder, valves and other rotating, sliding, moving parts.

- Store the machine in dry place duly protected from sun and rain.

Maintenance Job for Power Operated Equipment

All the above maintenance jobs apply to power equipment also. But the engines have to be taken care of specially. The life and efficiency of the engine mostly depends upon proper maintenance. For their running all engines need fuel, air and proper system of ignition. Thus in petrol engine, clean petrol, clean air and healthy ignition (spark plug & magnets) are essential. Besides those, the engine need perfect lubrication, too. In two stroke petrol engine, care must be taken to mix lubricating oil and petrol in exact ratio as recommended by engine manufacturer. Similarly in four stroke petrol engine the lubricating oil should be kept in sufficient quantity by observing the level gauge. The air cleaner should be cleaned occasionally. The spark plugs should be also cleaned, carbon removed and proper electrode gap should be maintained. The 2-stroke petrol engines used in low volume spraying should invariably be in good order otherwise the pesticide spraying will not be efficient.

Sufficient care should be taken at the depots to clean, oil and check equipment periodically when they are stored, and whenever machines are sent out to work, and when returned from field work. This minimum care to inspect the equipment, clean and flush and keep it duly oiled, would go a long way in improving the availability of good working sprayers and dusters and also prolonging their useful life.

Repairs and Replacements

The plant protection equipment is often found requiring frequent repairs and replacements which are both minor and major in nature. Due to this, a good number are found sick in the depots.

Hand operated equipment generally need minor repairs such as replacement of plunger washers, springs, nozzle etc., and these repair could as well be attended to by the operators themselves with little training and experience. It is essential to supply them necessary spare parts and tools well in time for repairing. In the case of power operated sprayers the engine repairs are classified into minor and major ones.

- Minor repairs: Spark plug cleaning and adjustment, air cleaner, carburetor cleaning, fuel cock and lines cleaning and starter repairs, etc. These can be attended to by the operators themselves with little experience and training.

- Major repairs: These repairs include replacement of parts like piston, rings, liners, crankshaft, bearings, valves, etc. These repairs have to be carried out systematically in well-equipped workshops by the competent and trained mechanics. Untrained personnel should not be allowed to handle such major repairs.

Suggestions on Maintenance

In order to improve the present situation the following suggestions are made:

- Plant protection equipment manufacturers, their dealers, State agril. Engineering workshops and extension officers need better coordination & cooperation to reduce the number of sick equipment.

- The field operating staff needs orientation training to be given on maintenance, repairs, operations and calibration of equipment on periodic basis.

- Adequate number of mechanics and supervisory staffs has to be posted for maintenance and repairs of the equipment.

- A district-wise service station, properly equipped, could cater to major repairs on power operated equipment within its zone.

PRINCIPLES OF PESTICIDE MANAGEMENT

Pesticide Management is the regulation of the import, manufacture, export, sale, transport, distribution, quality and use of pesticides with a view to:

- Control pests;

- Ensure availability of quality pesticides;

- Allow its use only after assessing its efficacy and safety;

- Minimize the contamination of agricultural commodities by pesticide residues;

- Create awareness among users regarding safe and judicious use of pesticides;

- To take necessary measures to continue, restrict or prohibit the use of pesticides on the basis of reassessment with a view to prevent its risk on human beings, animals or environment, and for matters connected there with or incidental.

Pesticide Management is an activity carried out within the overall framework of the Plant Production and Protection Division of FAO. It is designed to work together with member countries and other International Organizations as a partner to introduce sustainable and environmentally sound agricultural practices that reduce health and environmental risks associated with the use of pesticides.

In March 2007 FAO and WHO signed a Memorandum of Understanding on cooperation in a Joint Programme for the Sound Management of Pesticides to provide unified, coordinated and consistent advice and support to their Member States and to other stakeholders on sound management of pesticides. The "FAO/WHO Joint Meeting on Pesticide Management" (JMPM) is an expert ad hoc body administered jointly by FAO and WHO: the JMPM advises on matters pertaining to pesticide regulation, management and use, and alerts to new developments, problems or issues that otherwise merit attention from one or both Organizations. The JMPM consists of members drawn from the FAO Panel of Experts on Pesticide Management and the WHO Panel of Experts on Vector Biology and Control, which are statutory advisory bodies of the respective Organizations.

The International Code of Conduct on the Distribution and Use of Pesticides

The International Code of Conduct on the Distribution and Use of Pesticides is the worldwide guidance document on pesticide management for all public and private entities engaged in, or associated with, the distribution and use of pesticides. It was adopted for the first time in 1985 by the Twenty-fifth Session of the FAO Conference.

It focuses on:

- Risk reduction,

- Protection of human health and the environmental,

- Support for sustainable agricultural development by using pesticides in an effective manner and applying IPM strategies.

Particular concerns are given for countries where living and working conditions make pesticide use more risky.

The Code is designed to provide standards of conduct and to serve as a point of reference in relation to sound pesticide management practices, in particular for government authorities and the pesticide industry. Following the adoption of the Rotterdam Convention in 1998 and in view of the changing international policy framework, as well as the persistence of certain pesticide management problems, particularly in developing countries, in 1999 FAO initiated the update and revision process of the Code.

The 12 Articles of the Code, plus supporting technical guidelines and a new Annex consisting of references to international policy instruments related to the Code, represent an up-to-date standard for pesticide management. This embodies a modern approach, leading to sound management of pesticides which focuses on risk reduction, protection of human and environmental health, and support for sustainable agricultural development by using pesticides in an effective manner and applying IPM strategies.

In addition, the revised Code includes the life-cycle concept of pesticide management.

Objectives of the Code

The objectives of this Code are to establish voluntary standards of conduct for all public and private entities engaged in or associated with the distribution and use of pesticides, particularly where there is inadequate or no national legislation to regulate pesticides.

The Code describes the shared responsibility of many sectors of society to work together so that the benefits to be derived from the necessary and acceptable use of pesticides are achieved without significant adverse effects on human health or the environment.

The Code addresses the need for a cooperative effort between governments of pesticide exporting and importing countries to promote practices that minimize potential health and environmental risks associated with pesticides, while ensuring their effective use.

The entities which are addressed by this Code include international organizations, governments of exporting and importing countries, pesticide industry, application equipment industry, traders, food industry, users, and public-sector organizations such as environmental groups, consumer groups and trade unions.

The Code recognizes that training at all appropriate levels is an essential requirement in implementing and observing its provisions. Therefore, governments, pesticide industry, users of pesticides, international organizations, non-governmental organizations (NGOs) and other parties concerned should give high priority to training activities related to each Article of the Code.

The standards of conduct set forth in this Code:

- Encourage responsible and generally accepted trade practices.

- Assist countries which have not yet established regulatory controls on the quality and suitability of pesticide products needed in that country to promote the judicious and efficient use of such products and address the potential risks associated with their use.

- Promote practices which reduce risks in the handling of pesticides, including minimizing adverse effects on humans and the environment and preventing accidental poisoning resulting from improper handling.

- Ensure that pesticides are used effectively and efficiently for the improvement of agricultural production and of human, animal and plant health.

- Adopt the "life-cycle" concept to address all major aspects related to the development, regulation, production, management, packaging, labeling, distribution, handling, application, use and control, including post registration activities and disposal of all types of pesticides, including used pesticide containers.

- Promote Integrated Pest Management (IPM) (including integrated vector management for public health pests).

- Include reference to participation in information exchange and international agreements identified, in particular the Rotterdam Convention on the Prior Informed Consent Procedure for Certain Hazardous Chemicals and Pesticides in International Trade.

Pesticide Management

- Governments have the overall responsibility to regulate the availability, distribution and use of pesticides in their countries and should ensure the allocation of adequate resources for this mandate.

- Pesticide industry should adhere to the provisions of this Code as a standard for the manufacture, distribution and advertising of pesticides, particularly in countries lacking appropriate legislation and advisory services.

- Governments of pesticide exporting countries should, to the extent possible:

  ◦ Provide technical assistance to other countries, especially those lacking technical expertise in the assessment of the relevant data on pesticides.

  ◦ Ensure that good trading practices are followed in the export of pesticides, especially to those countries with limited or no regulatory schemes.

- Pesticide industry and traders should observe the following practices in pesticide management, especially in countries without legislation or means of implementing regulations:

  ○ Supply only pesticides of adequate quality, packaged and labeled as appropriate for each specific market.

  ○ In close cooperation with procurers of pesticides, adhere closely to provisions of FAO guidelines on tender procedures.

  ○ Pay special attention to the choice of pesticide formulations and to presentation, packaging and labeling in order to reduce risks to users and minimize adverse effects on the environment.

  ○ Provide, with each package of pesticide, information and instructions in a form and language adequate to ensure effective use and reduce risks during handling.

  ○ Be capable of providing effective technical support, backed up by full product stewardship to field level, including advice on disposal of pesticides and used pesticide containers, if necessary.

  ○ Retain an active interest in following their products to the end-user, keeping track of major uses and the occurrence of any problems arising from the use of their products, as a basis for determining the need for changes in labeling, directions for use, packaging, formulation or product availability.

  ○ Pesticides whose handling and application require the use of personal protective equipment that is uncomfortable, expensive or not readily available should be avoided, especially in the case of small-scale users in tropical climates. Preference should be given to pesticides that require inexpensive personal protective and application equipment and to procedures appropriate to the conditions under which the pesticides are to be handled and used.

- National and international organizations, governments and pesticide industry should take coordinated action to disseminate educational materials of all types to pesticide users, farmers, farmer organizations, agricultural workers, unions and other interested parties. Similarly, users should seek and understand educational materials before applying pesticides and should follow proper procedures.

- Concerted efforts should be made by governments to develop and promote the use of IPM. Furthermore, lending institutions, donor agencies and governments should support the development of national IPM policies and improved IPM concepts and practices. These should be based on scientific and other strategies that promote increased participation of farmers (including women's groups), extension agents and on-farm researchers.

- All stakeholders, including farmers and farmer associations, IPM researchers, extension agents, crop consultants, food industry, manufacturers of biological and chemical pesticides and application equipment, environmentalists and representatives of consumer groups should play a proactive role in the development and promotion of IPM.

- Governments, with the support of relevant international and regional organizations, should encourage and promote research on, and the development of, alternatives posing fewer risks: biological control agents and techniques, non-chemical pesticides and pesticides that are, as far as possible or desirable, target-specific, that degrade into innocuous constituent parts or metabolites after use and are of low risk to humans and the environment.

- Governments and the application equipment industry should develop and promote the use of pesticide application methods and equipment that pose low risks to human health and the environment and that are more efficient and cost-effective, and should conduct ongoing practical training in such activities.

- Governments, pesticide industry and national and international organizations should collaborate in developing and promoting resistance management strategies to prolong the useful life of valuable pesticides and reduce the adverse effects resulting from the development of resistance of pests to pesticides.

Testing of Pesticides

Pesticide industry should:

- Ensure that each pesticide and pesticide product is adequately and effectively tested by recognized procedures and test methods so as to fully evaluate its efficacy, behavior, fate, hazard and risk with regard to the various anticipated conditions in regions or countries of use.

- Ensure that such tests are conducted in accordance with sound scientific procedures and the principles of good laboratory practice.

- Ensure that the proposed use pattern, label claims and directions, packages, technical literature and advertising truly reflect the outcome of these scientific tests and assessments.

- Provide at the request of a country, methods for the analysis of any active ingredient or formulation that they manufacture, and provide the necessary analytical standards.

- Provide advice and assistance in the training of technical staff involved in the relevant analytical work. Formulators should actively support this effort.

- Conduct residue trials prior to marketing, at least in accordance with Codex.

Reducing Health and Environmental Risks

Governments should:

- Implement a pesticide registration and control system.

- Periodically review the pesticides marketed in their country, their acceptable uses and their availability to each sector of the public, and conduct special reviews when indicated by scientific evidence.

- Carry out health surveillance programmes of those who are occupationally exposed to pesticides and investigate, as well as document, poisoning cases.

- Provide guidance and instructions to health workers, physicians and hospital staff on the treatment of suspected pesticide poisoning.

- Establish national or regional poisoning information and control centres at strategic locations to provide immediate guidance on first aid and medical treatment, accessible at all times.

- Provide extension and advisory services and farmers' organizations with adequate information about practical IPM strategies and methods, as well as the range of pesticide products available for use.

- Implement a programme to monitor pesticide residues in food and the environment.

Pesticide industry should:

- Cooperate & provide poison-control centres and medical practitioners with information about pesticide hazards and on suitable treatment of pesticide poisoning.

- Make every reasonable effort to reduce risks posed by pesticides by:

  ○ Making less toxic formulations available.

  ○ Introducing products in ready-to-use packages.

  ○ Developing application methods and equipment that minimize exposure to pesticides.

  ○ Using returnable and refillable containers where effective container collection systems are in place.

  ○ Using containers that are not attractive for subsequent reuse and promoting programmes to discourage their reuse, where effective container collection systems are not in place.

  ○ Using containers that are not attractive to or easily opened by children, particularly for domestic use products.

  ○ Using clear and concise labeling.

- Halt sale and recall products when handling or use pose an unacceptable risk under any use directions or restrictions.

Government and industry should cooperate in further reducing risks by:

- Promoting the use of proper and affordable personal protective equipment.

- Making provisions for safe storage of pesticides at both warehouse and farm level.

- Establishing services to collect and safely dispose of used containers and small quantities of left-over pesticides.

- Protecting biodiversity and minimizing adverse effects of pesticides on the environment (water, soil and air) and on non-target organisms.

To avoid unjustified confusion and alarm among the public, concerned parties should consider all available facts and should promote responsible information dissemination on pesticides and their uses.

Regulatory and Technical Requirements

Governments should:

- Introduce the necessary legislation for the regulation of pesticides.

- Conduct risk evaluations and make risk management decisions based on all available data or information, as part of the registration process.

- Improve regulations in relation to collecting and recording data on import, export, manufacture, formulation, quality and quantity of pesticides.

- Permit pesticide application and personal protective equipment to be marketed only if they comply with established standards.

- Detect and control illegal trade in pesticides.

- When importing food and agricultural commodities, recognize good agricultural practices in countries with which they trade.

Availability and Use

Governments should use (where appropriate) the WHO classification of pesticides by hazard as the basis for their regulatory measures and associate the hazard class with well-recognized hazard symbols. When determining the risk and degree of restriction, the type of formulation and method of application should be taken into account.

Two methods of restricting availability can be exercised by the responsible authority: not registering a product or, as a condition of registration, restricting the availability to certain groups of users in accordance with a national assessment of the hazards involved in the use of the product.

Prohibition of the importation, sale and purchase of highly toxic and hazardous products, such as those included in WHO classes Ia and Ib, may be desirable if other control measures or good marketing practices are insufficient to ensure that the product can be handled with acceptable risk to the user.

Distribution and Trade

Governments should:

- Develop regulations and implement licensing procedures to ensure that, those involved in the sale of pesticides are capable of providing buyers with sound advice on risk reduction and efficient use.

- Take the necessary regulatory measures to prohibit the repackaging or decanting of any pesticide into food or beverage containers and rigidly enforce punitive measures that effectively deter such practices.

- Encourage a market-driven supply process to reduce the potential for accumulation of excessive stocks.

Pesticide industry should:

- Take all necessary steps to ensure that pesticides entering international trade conform at least to relevant FAO, WHO or equivalent specifications.

- Ensure that pesticides manufactured for export are subject to the same quality requirements and standards as those applied to comparable domestic products.

- Encourage importing agencies, national or regional formulators and their respective trade organizations to cooperate in order to achieve fair practices as well as marketing and distribution practices that reduce the risks posed by pesticides.

- Recognize that a pesticide may need to be recalled by a manufacturer and distributor when its use, as recommended, represents an unacceptable risk to human and animal health or the environment, and act accordingly.

- Endeavour to ensure that pesticides are traded by and purchased from reputable traders, who should preferably be members of a recognized trade organization.

- Ensure that persons involved in the sale of pesticides are trained adequately, hold appropriate government licenses and have access to sufficient information, such as material safety data sheets, so that they are capable of providing buyers with advice on risk reduction and efficient use.

- Provide, consistent with national requirements, a range of pack sizes and types that are appropriate for the needs of small-scale farmers and other local users, in order to reduce risks and to discourage sellers from repackaging products in unlabeled or inappropriate containers.

The procurer (government authority, growers' association, or individual farmer) should establish purchasing procedures to prevent the oversupply of pesticides and consider including requirements relating to extended pesticide storage, distribution and disposal services in a purchasing contract.

Information Exchange

Governments should:

- Promote the establishment or strengthening of networks for information exchange on pesticides through national institutions, international, regional and sub-regional organizations and public sector groups.

- Facilitate the exchange of information between regulatory authorities to strengthen cooperative efforts. The information to be exchanged should include:

  ◦ Actions to ban or severely restrict a pesticide in order to protect human health or the environment, and additional information upon request.

   ◦ Scientific, technical, economic, regulatory and legal information concerning pesticides including toxicological, environmental and safety data.

   ◦ The availability of resources and expertise associated with pesticide regulatory activities.

In addition, governments are encouraged to develop:

- Legislation and regulations that permit the provision of information to the public about pesticide risks and the regulatory process.

- Administrative procedures to provide transparency and facilitate the participation of the public in the regulatory process.

International organizations should provide information on specific pesticides (including guidance on methods of analysis) through the provision of criteria documents, fact sheets, training and other appropriate means.

Labeling, Packaging, Storage and Disposal

- All pesticide containers should be clearly labeled in accordance with applicable guidelines, at least in line with the FAO guidelines on good labeling practice.

- Industry should use labels that:

   ◦ Comply with registration requirements and include recommendations.

   ◦ Include appropriate symbols and pictograms whenever possible, in addition to written instructions, warnings and precautions in the appropriate language or languages.

   ◦ Include, in the appropriate language or languages, a warning against the reuse of containers and instructions for the safe disposal or decontamination of used containers.

   ◦ Clearly show the release date (month and year) of the lot or batch and contain relevant information on the storage stability of the product.

- Pesticide industry, in cooperation with government, should ensure that:

   ◦ Packaging, storage and disposal of pesticides conform to the relevant FAO, UNEP9, WHO guidelines or regulations or to other international guidelines, where applicable.

   ◦ Packaging or repackaging is carried out only on licensed premises where the responsible authority is satisfied that staff are adequately protected against toxic hazards, that the resulting product will be properly packaged and labeled, and that the content will conform to the relevant quality standards.

- Governments should take the necessary regulatory measures to prohibit the repackaging or decanting of any pesticide into food or beverage containers and rigidly enforce punitive measures that effectively deter such practices.

- Governments, with the help of pesticide industry should inventory obsolete or unusable stocks of pesticides and used containers, establish and implement an action plan for their disposal, or remediation in the case of contaminated sites, and record these activities.

- Pesticide industry should be encouraged, with multilateral cooperation, to assist in disposing of any banned or obsolete pesticides and of used containers, in an environmentally sound manner, including reuse with minimal risk where approved and appropriate.

- Governments, pesticide industry, international organizations and the agricultural community should implement policies and practices to prevent the accumulation of obsolete pesticides and used containers.

Advertising

- Governments should control, by means of legislation, the advertising of pesticides in all media to ensure that it is not in conflict with label directions and precautions.

- Pesticide industry should ensure that:

  - All statements used in advertising are technically justified.

  - Advertisements do not contain any statement or visual presentation which is likely to mislead the buyer, in particular with regard to the "safety" of the product, its nature, composition or suitability for use, official recognition or approval.

  - Pesticides which are legally restricted to use by trained or registered operators are not publicly advertised through journals other than those catering for such operators, unless the restricted availability is clearly and prominently shown.

  - No company or individual in any one country simultaneously markets different pesticide active ingredients or combinations of ingredients under a single brand name.

  - Advertisements do not misuse research results, quotations from technical and scientific literature or scientific jargon to make claims appear to have a scientific basis they do not possess.

  - Claims as to safety, including statements such as "safe", "non-poisonous", "harmless", "non-toxic" or "compatible with IPM," are not made, with or without a qualifying phrase such as "when used as directed".

  - Statements comparing the risk, hazard or "safety" of different pesticides or other substances are not made.

  - Misleading statements are not made concerning the effectiveness of the product.

  - Advertisements do not contain any visual representation of potentially dangerous practices, such as mixing or application without sufficient protective clothing, use near food or use by or in the vicinity of children.

  - Advertising or promotional material draws attention to the appropriate warning phrases and symbols as laid down in the FAO labeling guidelines.

  - Technical literature provides adequate information on correct practices, including the observance of recommended application rates, frequency of applications and pre-harvest intervals.

- All staff involved in sales promotion are adequately trained and possess sufficient technical knowledge to present complete, accurate and valid information on the products sold.

- Advertisements and promotional activities should not include inappropriate incentives or gifts to encourage the purchase of pesticides.

Monitoring and Observance of the Code

- The Code should be published and should be observed through collaborative action on the part of governments, individually or in regional groupings, appropriate organizations and bodies of the United Nations system, international, governmental and nongovernmental organizations and the pesticide industry.

- The Code should be brought to the attention of all concerned in the regulation, manufacture, distribution and use of pesticides, so that governments, individually or in regional groupings, pesticide industry, international institutions, pesticide user organizations, agricultural commodity industries and food industry groups (such as supermarkets) that are in a position to influence good agricultural practices, understand their shared responsibilities in working together to ensure that the objectives of the Code are achieved.

- All parties should observe this Code and should promote the principles and ethics expressed by the Code, irrespective of other parties' ability to observe the Code. Pesticide industry should cooperate fully in the observance of the Code and promote the principles and ethics expressed by the Code, irrespective of a government's ability to observe the Code.

- Independently of any measures taken with respect to the observance of this Code, all relevant legal rules, whether legislative, administrative, judicial or customary, dealing with liability, consumer protection, conservation, pollution control and other related subjects, should be strictly applied.

- FAO and other competent international organizations should give full support to the observance of the Code.

- Governments, in collaboration with FAO, should monitor the observance of the Code and report on progress made to the Director-General of FAO.

- NGOs and other interested parties are invited to monitor activities related to the implementation of the Code and report these to the Director-General of FAO.

- Governing Bodies of FAO should periodically review the relevance and effectiveness of the Code. The Code should be considered a dynamic text which must be brought up to date as required, taking into account technical, economic and social progress.

CHLORPYRIFOS

Chlorpyrifos (CPS) is an organophosphate pesticide used on crops, animals, and buildings, and in other settings, to kill a number of pests, including insects and worms. It acts on the nervous systems of insects by inhibiting the acetylcholinesterase enzyme. First developed as a nerve gas in World War II, chlorpyrifos was introduced in 1965 by Dow Chemical Company.

Chlorpyrifos is considered moderately hazardous to humans by the World Health Organization based on its acute toxicity. Exposure surpassing recommended levels has been linked to neurological effects, persistent developmental disorders, and autoimmune disorders. Exposure during pregnancy may harm the mental development of children, and most home uses of chlorpyrifos were banned in the U.S. in 2001. In agriculture, it is "one of the most widely used organophosphate insecticides" in the United States, and before being phased out for residential use it was one of the most used residential insecticides.

On March 29, 2017, EPA Administrator Scott Pruitt denied a petition to ban chlorpyrifos. However, on August 9, 2018, the U.S. 9th Circuit Court of Appeals ordered the EPA to ban the sale of chlorpyrifos in the United States within 60 days, though this ruling was almost immediately appealed by Trump administration lawyers.

On May 2019 the California Department of Pesticide Regulation announced it will "cancel the registration that allows chlorpyrifos to be sold in California", a process that could however take up to two years.

Uses

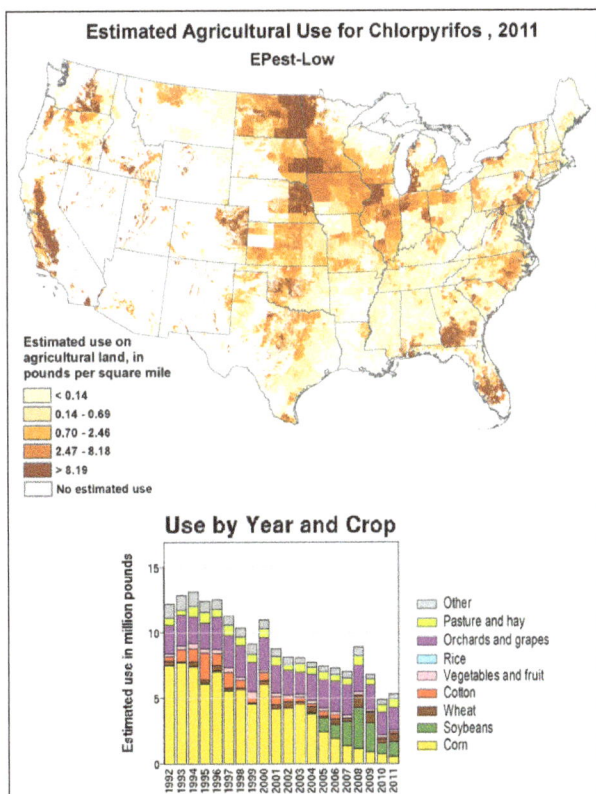

Chlorpyrifos is used in about 100 countriesaround the world to control insects in agricultural, residential, and commercial settings. Its use in residential applications is restricted in multiple countries. According to Dow, chlorpyrifos is registered for use in nearly 100 countries and is annually applied to approximately 8.5 million crop acres. The crops with the most use include cotton, corn, almonds, and fruit trees, including oranges, bananas, and apples.

Chlorpyrifos was first registered for use in the United States in 1965 for control of foliage and soil-born insects. The chemical became widely used in residential settings, on golf course turf, as a structural termite control agent, and in agriculture. Most residential use of chlorpyrifos has been phased out in the United States; however, agricultural use remains common.

EPA estimated that, between 1987 and 1998, about 21 million pounds of chlorpyrifos were used annually in the US. In 2001, chlorpyrifos ranked 15th among pesticides used in the United States, with an estimated 8 to 11 million pounds applied. In 2007, it ranked 14th among pesticide ingredients used in agriculture in the United States.

Application

Chlorpyrifos is normally supplied as a 23.5% or 50% liquid concentrate. The recommended concentration for direct-spray pin point application is 0.5% and for wide area application a 0.03 – 0.12% mix is recommended (US).

Toxicity

Chlorpyrifos exposure may lead to acute toxicity at higher doses. Persistent health effects follow acute poisoning or from long-term exposure to low doses, and developmental effects appear in fetuses and children even at very small doses.

Acute Health Effects

For acute effects, the World Health Organization classifies chlorpyrifos as Class II: moderately hazardous. The oral LD50 in experimental animals is 32 to 1000 mg/kg. The dermal LD50 in rats is greater than 2000 mg/kg and 1000 to 2000 mg/kg in rabbits. The 4-hour inhalation LC50 for chlorpyrifos in rats is greater than 200 mg/m^3.

Symptoms of Acute Exposure

Acute poisoning results mainly from interference with the acetylcholine neurotransmission pathway, leading to a range of neuromuscular symptoms. Relatively mild poisoning can result in eye watering, increased saliva and sweating, nausea and headache. Intermediate exposure may lead to muscle spasms or weakness, vomiting or diarrhea and impaired vision. Symptoms of severe poisoning include seizures, unconsciousness, paralysis, and suffocation from lung failure.

Children are more likely to experience muscle weakness rather than twitching; excessive saliva rather than sweat or tears; seizures; and sleepiness or coma.

Frequency of Acute Exposure

Acute poisoning is probably most common in agricultural areas in Asia, where many small farmers are affected. Poisoning may be due to occupational or accidental exposure or intentional self-harm.

Precise numbers of chlorpyrifos poisonings globally are not available. Pesticides are used in an estimated 200,000+ suicides annually with tens of thousands due to chlorpyrifos. Organophosphates are thought to constitute two thirds of ingested pesticides in rural Asia. Chlorpyrifos is among the commonly used pesticides used for self-harm.

In the US, the number of incidents of chlorpyrifos exposure reported to the US National Pesticide Information Center shrank sharply from over 200 in the year 2000 to less than 50 in 2003, following the residential ban.

Treatment

Poisoning is treated with atropine and simultaneously with oximes such as pralidoxime. Atropine blocks acetylcholine from binding with muscarinic receptors, which reduces the pesticide's impact. However, atropine does not affect acetylcholine at nicotinic receptors and thus is a partial treatment. Pralidoxime is intended to reactivate acetylcholinesterase, but the benefit of oxime treatment is questioned. A randomized controlled trial (RCT) supported the use of higher doses of pralidoxime rather than lower doses. A subsequent double-blind RCT, that treated patients who self-poisoned, found no benefit of pralidoxime, including specifically in chlorpyrifos patients.

Tourist Deaths

Chlorpyrifos poisoning was described by New Zealand scientists as the likely cause of death of several tourists in Chiang Mai, Thailand who developed myocarditis in 2011. Thai investigators came to no conclusion on the subject, but maintain that chlorpyrifos was not responsible and that the deaths were not linked.

Long Term Development

Epidemiological and experimental animal studies suggest that infants and children are more susceptible than adults to effects from low dose exposure. Chlorpyrifos has been suggested to have negative impacts on the cognitive functions in the developing brain. The young have a decreased capacity to detoxify chlorpyrifos and its metabolites. It is suggested that adolescents differ from adults in the metabolism of these compounds due to the maturation of organs in adolescents. This results in disruption in nervous system developmental processes, as observed in animal experiments. There are a number of studies observed in animals that show that chlorpyrifos alters the expression of essential genes that assist in the development of the brain.

1. Human studies: In multiple epidemiological studies, chlorpyrifos exposure during gestation or childhood has been linked with lower birth weight and neurological changes such as slower motor development and attention problems. Children with prenatal exposures to chlorpyrifos have been shown to have lower IQs. They have also been shown to have a higher chance of developing autism, attention deficit problems, and developmental disorders. A cohort of 7-year-old children was studied for neurological damages from prenatal exposure to chlorpyrifos. The study determined that the exposed children had deficits in working memory and full scale intelligence quotient (IQ). In a study on groups of Chinese infants, those exposed to chlorpyrifos showed significant decreases in motor functions such as reflexes, locomotion, and grasping at 9 months compared to those not exposed. Exposure to organophosphate pesticides in general has been increasingly associated with

changes in children's cognitive, behavioral and motor performance. Infant girls were shown to be more susceptible to harmful effects from organophosphate insecticides than infant boys.

2. Animal experiments: In experiments with rats, early, short-term low-dose exposure to chlorpyrifos resulted in lasting neurological changes, with larger effects on emotional processing and cognition than on motor skills. Such rats exhibited behaviors consistent with depression and reduced anxiety. In rats, low-level exposure during development has its greatest neurotoxic effects during the period in which sex differences in the brain develop. Exposure leads to reductions or reversals of normal gender differences. Exposure to low levels of chlorpyrifos early in rat life or as adults also affects metabolism and body weight. These rats show increased body weight as well as changes in liver function and chemical indicators similar to prediabetes, likely associated with changes to the cyclic AMP system. Moreover, experiments with zebra fish showed significant detriments to survivability, reproductive processes, and motor function. Varying doses created a 30%-100% mortality rate of embryos after 90 days. Embryos were shown to have decreased mitosis, resulting in mortality or developmental dysfunctions. In the experiments where embryos did survive, spinal lordosis and lower motor functions were observed. The same study showed that chlorpyrifos had more severe morphological deformities and mortality in embryos than diazinon, another commonly used organophosphate insecticide.

Adulthood

Adults may develop lingering health effects following acute exposure or repeated low-dose exposure. Among agricultural workers, chlorpyrifos has been associated with slightly increased risk of wheeze, a whistling sound while breathing due to obstruction of the airways.

Among 50 farm pesticides studied, chlorpyrifos was associated with higher risks of lung cancer among frequent pesticide applicators than among infrequent or non-users. Pesticide applicators as a whole were found to have a 50% lower cancer risk than the general public, likely due to their nearly 50% lower smoking rate. However, chlorpyrifos applicators had a 15% lower cancer risk than the general public, which the study suggests indicates a link between chlorpyrifos application and lung cancer.

Twelve people who had been exposed to chlorpyrifos were studied over periods of 1 to 4.5 years. They were found to have heightened immune responses to common allergens and increased antibiotic sensitivities, elevated CD26 cells, and a higher rate of autoimmunity, compared with control groups. Autoantibodies were directed toward smooth muscle, parietal cell, brush border, thyroid gland, myelin, and the subjects also had more anti-nuclear antibodies.

Mechanisms of Toxicity

Acetylcholine Neurotransmission

Primarily, chlorpyrifos and other organophosphate pesticides interfere with signaling from the neurotransmitter acetylcholine. One chlorpyrifos metabolite, chlorpyrifos-oxon, binds permanently to the enzyme acetylcholinesterase, preventing this enzyme from deactivating acetylcholine in the synapse. By irreversibly inhibiting acetylcholinesterase, chlorpyrifos leads to a build-up of acetylcholine between neurons and a stronger, longer-lasting signal to the next neuron. Only when new molecules of acetylcholinesterase have been synthesized can normal function return. Acute symptoms of chlorpyrifos poisoning only occur when more than 70% of acetylcholinesterase molecules are inhibited. This mechanism is well established for acute chlorpyrifos poisoning and also some lower-dose health impacts. It is also the primary insecticidal mechanism.

Non-cholinesterase Mechanisms

Chlorpyrifos may affect other neurotransmitters, enzymes and cell signaling pathways, potentially at doses below those that substantially inhibit acetylcholinesterase. The extent of and mechanisms for these effects remain to be fully characterized. Laboratory experiments in rats and cell cultures suggest that exposure to low doses of chlorpyrifos may alter serotonin signaling and increase rat symptoms of depression; change the expression or activity of several serine hydrolase enzymes, including neuropathy target esterase and several endocannabinoid enzymes; affect components of the cyclic AMP system; and influence other chemical pathways.

Paraoxonase Activity

The enzyme paraoxonase 1 (PON1) detoxifies chlorpyrifos oxon, the more toxic metabolite of chlorpyrifos, via hydrolysis. In laboratory animals, additional PON1 protects against chlorpyrifos toxicity while individuals that do not produce PON1 are particularly susceptible. In humans, studies about the effect of PON1 activity on the toxicity of chlorpyrifos and other organophosphates are mixed, with modest yet inconclusive evidence that higher levels of PON1 activity may protect against chlorpyrifos exposure in adults; PON1 activity may be most likely to offer protection from low-level chronic doses. Human populations have genetic variation in the sequence of PON1 and its promoter region that may influence the effectiveness of PON1 at detoxifying chlorpyrifos oxon and the amount of PON1 available to do so. Some evidence indicates that children born to women with low PON1 may be particularly susceptible to chlorpyrifos exposure. Further, infants produce low levels of PON1 until six months to several years after birth, likely increasing the risk from chlorpyrifos exposure early in life.

Combined Exposures

Several studies have examined the effects of combined exposure to chlorpyrifos and other chemical agents, and these combined exposures can result in different effects during development. Female rats exposed first to dexamethasone, a treatment for premature labor, for three days in utero and then to low levels of chlorpyrifos for four days after birth experienced additional damage to the acetylcholine system upstream of the synapse that was not observed with either exposure alone. In both male and female rats, combined exposures to dexamethasone and chlorpyrifos decreased serotonin turnover in the synapse, for female rats with a greater-than-additive result. Rats that were co-exposed to dexamethasone and chlorpyrifos also exhibited complex behavioral differences from exposure to either chemical alone, including lessening or reversing normal sex differences in behavior. In the lab, in rats and neural cells co-exposed to both nicotine and chlorpyrifos, nicotine appears to protect against chlorpyrifos acetylcholinesterase inhibition and reduce its effects on neurodevelopment. In at least one study, nicotine appeared to enhance chlorpyrifos detoxification.

Human Exposure

In 2011, EPA estimated that, in the general US population, people consume 0.009 micrograms of chlorpyrifos per kilogram of their body weight per day directly from food residue. Children are estimated to consume a greater quantity of chlorpyrifos per unit of body weight from food residue, with toddlers the highest at 0.025 micrograms of chlorpyrifos per kilogram of their body weight per day. People may also ingest chlorpyrifos from drinking water or from residue in food handling establishments. The EPA's acceptable daily dose is 0.3 micrograms/kg/day. However, as of 2016, EPA scientists had not been able to find any level of exposure to the pesticide that was safe. The EPA 2016 report states in part "this assessment indicates that dietary risks from food alone are of concern". The report also states that previous published risk assessments for "chlorpyrifos may not provide a sufficiently health protective human health risk assessment given the potential for neurodevelopmental outcomes."

Humans can be exposed to chlorpyrifos by way of ingestion (e.g., residue on treated produce, drinking water), inhalation (especially of indoor air), or absorption (i.e., through the skin). However, compared to other organophosphates, chlorpyrifos degrades relatively quickly once released into the environment. According to the National Institutes of Health, the half-life for chlorpyrifos (i.e., the period of time that it takes for the active amount of the chemical to decrease by 50%) "can typically range from 33-56 days for soil incorporated applications and 7-15 days for surface applications"; in water, the half-life is about 25 days, and in the air, the half-life can range from four to ten days.

Before residential use was restricted in the US, data from 1999 to 2000 in the national NHANES study detected the metabolite TCPy in 91% of human urine samples tested. In samples collected between 2007 and 2009 from families living in Northern California, TCPy was found in 98.7% of floor wipes tested and in 65% of urine samples tested. For both children and adults, the average concentrations of TCPy in urine were lower in the later study. A study looking at pregnant women living in an agricultural community in the Salinas Valley, CA in 2004 showed that 76% of the pregnant women had detectable levels of TCPy. A 2008 study found dramatic drops in the urinary levels of chlorpyrifos metabolites when children in the general population switched from conventional to organic diets.

Children of agricultural workers are more likely to come into contact with chlorpyrifos. A study done in an agricultural community in Washington State showed that children who lived in closer proximity to farmlands had higher levels of chlorpyrifos residues from house dust. Chlorpyrifos residues were also found on work boots and children's hands, showing that agricultural families could take home these residues from their jobs. Urban and suburban children get most of their chlorpyrifos exposure from fruits and vegetables. A study done in North Carolina on children's exposure showed that chlorpyrifos was detected in 50% of the food, dust, and air samples in both their homes and daycare with the main route of exposure being through ingestion. Certain other populations with higher likely exposure to chlorpyrifos, such as people who apply pesticides, work on farms, or live in agricultural communities, have been measured in the US to excrete TCPy in their urine at levels that are 5 to 10 times greater than levels in the general population.

As of 2016, chlorpyrifos is the most used conventional insecticide in the US and is used in over 40 states; the top five states (in total pounds applied) are California, North Dakota, Minnesota, Iowa, and Texas. It is used on over 50 crops, with the top five crops (in total pounds applied) being soybeans, corn, alfalfa, oranges, and almonds. Additionally, crops with 30% or more of the crop treated (compared to total acres grown) include apples, asparagus, walnuts, table grapes, cherries, cauliflower, broccoli, and onions.

Air monitoring studies conducted by the California Air Resources Board (CARB) documented chlorpyrifos in the air of California communities. Analyses indicate that children living in areas of high chlorpyrifos use are often exposed to levels that exceed EPA dosages. A study done in Washington state using passive air samplers showed that households who lived less than 250 meters from a fruit tree field had higher levels of chlorpyrifos concentrations in the air than households that were further away. Advocacy groups monitored air samples in Washington and Lindsay, California, in 2006 with comparable results. Grower and pesticide industry groups argued that the air levels documented in these studies are not high enough to cause significant exposure or adverse effects, but a follow-up biomonitoring study in Lindsay showed that people there display above-normal chlorpyrifos levels.

Effects on Wildlife

Aquatic Life

Among freshwater aquatic organisms, crustaceans and insects appear to be more sensitive to acute exposure than are fish. Aquatic insects and animals appear to absorb chlorpyrifos directly from water rather than ingesting it with their diet or through sediment exposure.

Concentrated chlorpyrifos released into rivers killed insects, shrimp and fish. In Britain, the rivers Roding, Ouse, Wey, and Kennet all experienced insect, shrimp, and fish kills as a result of small releases of concentrated chlorpyrifos. The July 2013 release along the River Kennet poisoned insect life and shrimp along 15 km of the river, likely from a half a cup of concentrated chlorpyrifos washed down a drain.

Bees

Acute exposure to chlorpyrifos can be toxic to bees, with an oral LD50 of 360 ng/bee and a contact LD50 of 70 ng/bee. Guidelines for Washington state recommend that chlorpyrifos products should not be applied to flowering plants such as fruit trees within 4–6 days of blossoming to prevent bees from directly contacting the residue.

Risk assessments have primarily considered acute exposure, but more recently researchers have begun to investigate the effects of chronic, low-level exposure through residue in pollen and components of bee hives. A review of US studies, several European countries, Brazil and India found chlorpyrifos in nearly 15% of hive pollen samples and just over 20% of honey samples. Because of its high toxicity and prevalence in pollen and honey, bees are considered to have higher risk from chlorpyrifos exposure via their diet than from many other pesticides.

When exposed in the laboratory to chlorpyrifos at levels roughly estimated from measurements in hives, bee larvae experienced 60% mortality over 6 days, compared with 15% mortality in controls. Adult bees exposed to sub-lethal effects of chlorpyrifos (0.46 ng/bee) exhibited altered behaviors: less walking; more grooming, particularly of the head; more difficulty righting themselves; and unusual abdominal spasms. Chlorpyrifos oxon appears to particularly inhibit acetylcholinesterase in bee gut tissue as opposed to head tissue. Other organophosphate pesticides impaired bee learning and memory of smells in the laboratory.

Regulation

International Law

Chlorpyrifos is not regulated under international law or treaty. Organizations such as PANNA and the NRDC state that chlorpyrifos meets the four criteria (persistence, bioaccumulation, long-range transport, and toxicity) in Annex D of the Stockholm Convention on Persistent Organic Pollutants and should be restricted.

National Regulations

Chlorpyrifos was used to control insect infestations of homes and commercial buildings in Europe until it was banned from sale in 2008. Chlorpyrifos is restricted from termite control in Singapore as of 2009. It was banned from residential use in South Africa as of 2010. In 2010, India barred Dow from commercial activity for 5 years after India's Central Bureau of Investigation found Dow guilty of bribing Indian officials in 2007 to allow the sale of chlorpyrifos.

United States

In the United States, several laws directly or indirectly regulate the use of pesticides. These laws, which are implemented by the EPA, NIOSH, USDA and FDA, include: the Clean Water Act (CWA); the Endangered Species Act (ESA); the Federal Insecticide, Fungicide, and Rodenticide Act (FIFRA); the Federal Food, Drug, and Cosmetic Act (FFDCA); the Comprehensive Environmental Response, Compensation, and Liability Act (CERCLA); and the Emergency Planning and Community Right-to-Know Act (EPCRA). As a pesticide, chlorpyrifos is not regulated under the Toxic Substances Control Act (TSCA).

Chlorpyrifos is sold in restricted-use products for certified pesticide applicators to use in agriculture and other settings, such as golf courses or for mosquito control. It may also be sold in ant and roach baits with childproof packaging. In 2000, manufacturers reached an agreement with the EPA to voluntarily restrict the use of chlorpyrifos in places where children may be exposed, including homes, schools and day care centers.

In 2007 Pesticide Action Network North America and Natural Resources Defense Council (collectively, PANNA) submitted an administrative petition requesting a chlorpyrifos ban. On August 10, 2015, the Ninth Circuit Court of Appeals in PANNA v. EPA ordered the EPA to respond to PANNA's petition by "revoking all tolerances for the insecticide chlorpyrifos", denying the Petition or issuing a "proposed or final tolerance revocation" no later than October 31, 2015. The EPA was "unable to conclude that the risk from aggregate exposure from the use of chlorpyrifos [met] the safety standard of the Federal Food, Drug, and Cosmetic Act (FFDCA)" and therefore proposed "to revoke all tolerances for chlorpyrifos."

In an October 30, 2015 statement Dow AgroSciences disagreed with the EPA's proposed revocation and "remained confident that authorized uses of chlorpyrifos products, as directed, offer wide margins of protection for human health and safety." In a November 2016 press release, DOW argued that chlorpyrifos was "a critical tool for growers of more than 50 different types of crops in the United States" with limited or no viable alternatives. The Environment News Service quoted the Dow AgroSciences statement disagreeing with the EPA findings.

Chlorpyrifos is one of the most widely used pest control products in the world. It is authorized for use in about 100 nations, including the U.S., Canada, the United Kingdom, Spain, France, Italy, Japan, Australia and New Zealand, where it is registered for protection of essentially every crop now under cultivation. No other pesticide has been more thoroughly tested.

In November 2016, the EPA reassessed its ban proposal after taking into consideration recommendations made by the agency's Science Advisory Panel which had rejected the EPA's methodology in quantifying the risk posed by chlorpyrifos. Using a different methodology as suggested by the panel, the EPA retained its decision to completely ban chlorpyrifos. The EPA concluded that, while "uncertainties" remain, a number of studies provide "sufficient evidence" that children experience neurodevelopment effects even at low levels of chlorpyrifos exposure.

On March 29, 2017, EPA Administrator Scott Pruitt, appointed by the Trump administration, over-turned the 2015 EPA revocation and denied the administrative petition by the Natural Resources Defense Council and the Pesticide Action Network North America to ban chlorpyrifos.

By reversing the previous administration's steps to ban one of the most widely used pesticides in the world, we are returning to using sound science in decision-making – rather than predeter-mined results.

The American Academy of Pediatrics responded to the administration's decision saying they are "deeply alarmed" by Pruitt's decision to allow the pesticide's continued use. "There is a wealth of science demonstrating the detrimental effects of chlorpyrifos exposure to developing fetuses, in-fants, children and pregnant women. The risk to infant and children's health and development is unambiguous."

Asked in April whether Pruitt had met with Dow Chemical Company executives or lobbyists before his decision, an EPA spokesman replied: "We have had no meetings with Dow on this topic." In June, after several Freedom of Information Act requests, the EPA released a copy of Pruitt's March meeting schedule which showed that a meeting had been scheduled between Pruitt and Dow CEO Andrew Liveris at a hotel in Houston, Texas, on March 9. Both men were featured speakers at an energy conference.

In August, it was revealed that in fact Pruitt and other EPA officials had met with industry repre-sentatives on dozens of occasions in the weeks immediately prior to the March decision, promis-ing them that it was "a new day" and assuring them that their wish to continue using chlorpyrifos had been heard. Ryan Jackson, Pruitt's chief of staff, said in a March 8 email that he had "scared" career staff into going along with the political decision to deny the ban, adding "They know where this is headed and they are documenting it well."

On August 9, 2018 the U.S. 9th Circuit court of Appeals ruled that the EPA must ban chlorpyrifos within 60 days from that date. A spokesman for Dow DuPont stated that "all appellate options" would be considered. In contrast, Marisa Ordonia, a lawyer for Earthjustice, the organization that had conducted much of the legal work on the case, hailed the decision.

Residue

The use of chlorpyrifos in agriculture can leave chemical residue on food commodities. The FFDCA requires EPA to set limits, known as tolerances, for pesticide residue in human food and animal feed products based on risk quotients for acute and chronic exposure from food in humans. These tolerances limit the amount of chlorpyrifos that can be applied to crops. FDA enforces EPA's pesticide tolerances and determines "action levels" for the unintended drift of pesticide residues onto crops without tolerances.

After years of research without a conclusion and cognizant of the court order to issue a final ruling, the EPA proposed to eliminate all tolerances for chlorpyrifos ("Because tolerances are the maximum residue of a pesticide that can be in or on food, this proposed rule revoking all chlorpyrifos tolerances means that if this approach is finalized, all agricultural uses of chlorpyrifos would cease") and solicited comments.

The Dow Chemical Company is actively opposed to tolerance restrictions on chlropyrifos and is currently lobbying the White House to, among other goals, pressure EPA to reverse its proposal to revoke chlorpyrifos food residue tolerances.

The EPA has not updated the approximately 112 tolerances pertaining to food products and supplies since 2006. However, in a 2016 report, EPA scientists had not been able to find any level of exposure to the pesticide that was safe. The EPA 2016 report states in part "this assessment indicates that dietary risks from food alone are of concern" the report also states that previous published risk assessments for "chlorpyrifos may not provide a sufficient human health risk assessment given the potential for neurodevelopmental outcomes."

"The food only exposures for chlorpyrifos are of risk concern for all population subgroups analyzed. Children (1–2 years old) is the population subgroup with the highest risk estimate at 14,000% of the ssPAD food."

Based on 2006 EPA rules, Chlorpyrifos has a tolerance of 0.1 part per million (ppm) residue on all food items unless a different tolerance has been set for that item or chlorpyrifos is not registered for use on that crop. EPA set approximately 112 tolerances pertaining to food products and supplies. In 2006, to reduce childhood exposure, the EPA amended its chlorpyrifos tolerance on apples, grapes and tomatoes, reducing the grape and apple tolerances to 0.01 ppm and eliminating the tolerance on tomatoes. Chlorpyrifos is not allowed on crops such as spinach, squash, carrots, and tomatoes; any chlorpyrifos residue on these crops normally represents chlorpyrifos misuse or spray drift.

Food handling establishments (places where food products are held, processed, prepared or served) are included in the food tolerance of 0.1 ppm for chlorpyrifos. Food handling establishments may use a 0.5% solution of chlorpyrifos solely for spot and crack and crevice treatments. Food items are to be removed or protected during treatment. Food handling establishment tolerances may be modified or exempted under FFDCA.

Water

Chlorpyrifos in waterways is regulated as a hazardous substance under Federal Water Pollution Control Act and falls under the CWA amendments of 1977 and 1978. The regulation is inclusive of all chlorpyrifos isomers and hydrates in any solution or mixture. EPA has not set a drinking water regulatory standard for chlorpyrifos, but has established a drinking water guideline of 2 ug/L.

In 2009, in order to protect threatened salmon and steelhead under CWA and ESA, EPA and National Marine Fisheries Service (NMFS) recommended limits on the use of chlorpyrifos in California, Idaho, Oregon and Washington and requested that manufacturers voluntarily add buffer zones, application limits and fish toxicity to the standard labeling requirements for all chlorpyrifos-based products. Manufacturers rejected the request. In February 2013 in Dow AgroSciences vs NMFS, the Fourth Circuit Court of Appeals vacated EPA's order for these labeling requirements. In August 2014, in the settlement of a suit brought by environmental and fisheries advocacy groups against EPA in the U.S. District Court for the Western District of Washington, EPA agreed to re-instate no-spray stream buffer zones in California, Oregon and Washington, restricting aerial spraying (300 ft.) and ground-based applications (60 ft.) near salmon populations. These buffers will remain until EPA makes a permanent decision in consultation with NMFS.

Reporting

EPCRA designates the chemicals that facilities must report to the Toxics Release Inventory (TRI), based on EPA assessments. Chlorpyrifos is not on the reporting list. It is on the list of hazardous substances under CERCLA (the Superfund Act). In the event of an environmental release above its reportable quantity of 1 lb or 0.454 kg, facilities are required to immediately notify the National Response Center (NRC).

In 1995, Dow paid a $732,000 EPA penalty for not forwarding reports it had received on 249 chlorpyrifos poisoning incidents.

Occupational Exposure

In 1989, OSHA established a workplace permissible exposure limit (PEL) of 0.2 mg/m^3 for chlorpyrifos, based on an 8-hour time weighted average (TWA) exposure. However, the rule was remanded by the U.S. Circuit Court of Appeals and no PELs are in place presently.

EPA's Worker Protection Standard requires owners and operators of agricultural businesses to comply with safety protocols for agricultural workers and pesticide handlers (those who mix, load and apply pesticides). For example, in 2005, the EPA filed an administrative complaint against JSH Farms, Inc. (Wapato, Washington) with proposed penalties of $1,680 for using chlorpyrifos in 2004 without proper equipment. An adjacent property was contaminated with chlorpyrifos due to pesticide drift and the property owner suffered from eye and skin irritation.

Manufacture

Chlorpyrifos is produced via a multistep synthesis from 3-methylpyridine, eventually reacting 3,5,6-trichloro-2-pyridinol with diethylthiophosphoryl chloride.

PYRIPROXYFEN

Pyriproxyfen is a pesticide which is found to be effective against a variety of insects. It was introduced to the US in 1996, to protect cotton crops against whitefly. It has also been found useful for protecting other crops. It is also used as prevention for flea control on household pets, for killing indoor and outdoor ants and roaches. Methods of application include aerosols, bait, carpet powders, foggers, shampoos and pet collars.

Pyriproxyfen is a juvenile hormone analog and an insect growth regulator. It prevents larvae from developing into adulthood and thus rendering them unable to reproduce.

In the US, pyriproxyfen is often marketed under the trade name *Nylar*. In Europe, pyriproxyfen is known under the brand names Cyclio (Virbac) and Exil Flea Free TwinSpot (Emax).

Toxicity in Mammals

Pyriproxyfen has low acute toxicity. According to WHO and FAO, at elevated doses exceeding 5000 mg/kg of body weight, pyriproxyfen affects the liver in mice, rats and dogs. It also changes cholesterol levels, and may cause modest anemia at high doses.

Rumor of Link to Microcephaly Outbreak in Brazil

Starting in 2014, pyriproxifen was put into Brazilian water supplies to fight the proliferation of mosquito larvae. This is in line with the World Health Organization (WHO)'s Pesticide Evaluation Scheme (WHOPES) for larvicides. In January 2016, the Brazilian Association for Collective Health criticized the introduction of pyriproxyfen in Brazil. Abrasco demanded the "immediate suspension of pyriproxyfen and all growth inhibitors in drinking water." The organization is opposed to the use of growth inhibitors in the context of an ongoing outbreak of fetal malformation.

On February 3, the rumor that pyriproxyfen, not the Zika virus, is the cause of the 2015-2016 microcephaly outbreak in Brazil was raised in a report of the Argentinean organization Physicians in the Crop-Sprayed Villages (PCST). It attracted wide media coverage. Abrasco clarified that position as an misinterpretation of their statement, saying "at no time did we state that pesticides, insecticides, or other chemicals are responsible for the increasing number of microcephaly cases in Brazil". They also condemned the behavior of the websites that spread the misinformation, adding that such "untruths violates the anguish and suffering of the people in vulnerable positions". In addition, the coordinator for the PCST statement, Medardo Ávila Vazquez, acknowledged in an interview that "the group hasn't done any lab studies or epidemiological research to support its assertions, but it argues that using larvicides may cause human deformities."

On February 13, the Brazilian state of Rio Grande do Sul suspended pyriproxyfen's use, citing both Abrasco and PCST positions. The Health Minister of Brazil, Marcelo Castro, criticized this step, noting that the claim is "a rumor lacking logic and sense. It has no basis." They also noted that the insecticide is approved by the National Sanitary Monitoring Agency and "all regulatory agencies in the whole world". The manufacturer of the insecticide, Sumitomo Chemical, stated "there is no scientific basis for such a claim" and also referred to the approval of pyriproxyfen by the World Health Organization since 2004 and the United States Environmental Protection Agency since 2001.

George Dimech, the director of Disease Control and Diseases of the Health Department of Pernambuco in Brazil, gave an interview to the BBC where he pointed out that the city of Recife has the current highest reported number of cases of microcephaly, yet pyriproxyfen is not used in the region, but another insecticide altogether. He added that "this lack of spatial correlation weakens the idea that the larvicide is the cause of the problem." In addition, the BBC interviewed researchers in Pernambuco, where no evidence has been found of the cases being linked to any environmental cause like an insecticide. Neurologist Vanessa van der Linden stated in an interview, "Clinically, the changes we see in the scans of babies suggest that the injuries were caused by congenital infection and not by larvicide, drug or vaccine."

Noted skeptic David Gorski called the claim a conspiracy theory and pointed out that antivaccine proponents had also claimed that the Tdap vaccine was the cause of the microcephaly outbreak, due to its introduction in 2014, along with adding, "One can't help but wonder what else the Brazilian Ministry of Health did in 2014 that cranks can blame microcephaly on." Gorski also pointed out the extensive physiochemical understanding of pyriproxyfen coded in the WHO Guidelines for Drinking-water Quality, which concluded in a past evaluation that the insecticide is not genotoxic, and that the doctor organization making the claim has been advocating against all pesticides since 2010, complicating their reliability.

A professor from the University of Adelaide in Australia, stated that "The effect of pyriproxyfen on reproduction and fetal abnormalities is well studied in animals. In a variety of animal species even enormous quantities of pyriproxyfen do not cause the defects seen during the recent Zika outbreak." A colleague also from the University of Adelaide stated that "While the evidence that Zika virus is responsible for the rise in microcephaly in Brazil is not conclusive, the role of pyriproxyfen is simply not plausible." Another professor in Australia concluded that "insect development is quite different to human development and involves different hormones, developmental pathways and sets of genes, so it cannot be assumed that chemicals affecting insect development also influence mammalian development."

ACARICIDE

Acaricides are pesticides that kill ticks and mites, closely related groups of invertebrates. They are one part of a strategy for controlling ticks around homes and should be combined with measures to reduce tick habitats.

An acaricide for tick control will include active ingredients like permethrin, cyfluthrin, bifenthrin, carbaryl, and pyrethrin. These chemicals are sometimes called acaricide insecticides, but ticks are arachnids, not insects, so this isn't technically accurate.

Some acaricides are available for homeowners to use. Others can only be sold to licensed applicators, so you'll need to hire a professional to apply them. Diatomaceous earth is a non-chemical alternative that may help to suppress tick populations.

Using an Acaricide

There are two main ways to use an acaricide for tick control. First, the acaricide can be applied to a whole area. Second, it can be used to treat the hosts that carry ticks, including rodents and deer.

The best time for an area-wide acaricide application is in mid-May through mid-June, when ticks are in the nymphal stage. Another application can be done in the fall to target adult ticks.

Acaricides can be applied to tick habitats around a residence including wooded areas and their borders, stone walls, and ornamental gardens. Using acaricides in lawns is only recommended when residential areas are located directly next to woodlands or include wooded sections.

To treat deer tick hosts, rodent bait boxes and deer feeding stations can be placed on a property. These devices attract the animals with food or nesting material, then dose them with an acaricide. The process is harmless to the animal and help can suppress tick populations in the area. Permits may be needed, so check with local authorities before setting them up.

Other ways to keep ticks away from the home include the following strategies:

- The deer tick mainly feeds on white-tailed deer and on rodents, so reducing the attractiveness of your yard for these critters can also reduce the tick population.

- Tall grass, brush, leaf piles, and debris all provide tick habitat, so keep grass mowed and remove brush around the home. Neatly stack wood, and consider eliminating stone walls and wood piles. Adding a 3-foot-wide strip of mulch or gravel can keep ticks from crossing into the garden from a nearby wooded area.

Terminology

More specific words are sometimes used, depending upon the targeted group:

- "Ixodicides" are substances that kill ticks.

- "Miticides" are substances that kill mites.

- The term scabicide is more narrow, and refers to agents specifically targeting *Sarcoptes*.

- The term "arachnicide" is more general, and refers to agents that target arachnids. This term is used much more rarely, but occasionally appears in informal writing.

As a practical matter, mites are a paraphyletic grouping, and mites and ticks are usually treated as a single group.

Examples include:

- Permethrin can be applied as a spray. The effects are not limited to mites: lice, cockroaches, fleas, mosquitos, and other insects will be affected.

- Ivermectin can be prescribed by a medical doctor to rid humans of mite and lice infestations, and agricultural formulations are available for infested birds and rodents.

- Antibiotic miticides.

- Carbamate miticides.

- Dienochlor miticides.

- Formamidine miticides.

- Organophosphate miticides.

- Diatomaceous earth will also kill mites by disrupting their cuticles, which dries out the mites.

- Dicofol, a compound structurally related to the insecticide DDT, is a miticide that is effective against the red spider mite *Tetranychus urticae.*

- Lime sulfur is effective against sarcoptic mange. It is made by mixing hydrated lime, sulfur, and water, and boiling for about 1 hour. Hydrated lime can bond with about 1.7 its weight of sulfur. (Quicklime can bond with as much as 2.2 times its weight of sulfur). The strongest concentrate is diluted 1:32 before saturating the skin (avoiding the eyes), applied at six-day intervals.

- Nonpesticide miticides act by causing desiccation, but are not a diatomaceous earth (which contain crystalline silica, potentially dangerous by inhalation), but made from a patented mix of food-grade components, one to breach the cuticle and one to ensure rapid, reliable desiccation. They can be dusted as powder or sprayed in aqueous solution.

- A variety of commercially available systemic and non-systemic miticides: abamectin, acequinocyl, bifenazate, bifenazate, chlorfenapyr, clofentezine, cyflumetofen, cypermethrin, dicofol, etoxazole, fenazaquin, fenpyroximate, hexythiazox, imidacloprid, propargite, pyridaben, spiromesifen, spirotetramat.

Acaricides are also being used in attempts to stop rhinoceros poaching. Holes are drilled into the horn of a sedated rhino and acaricide is pumped in and pressurized. Should the horn be consumed by humans as in traditional Chinese medicine, it is expected to cause nausea, stomachache, and diarrhea, or convulsions, depending on the quantity, but not fatalities. Signs posted at wildlife refuges that the rhinos therein have been treated are thus expected to deter poaching. The original idea grew out of research into using the horn as a reservoir for one-time tick treatments; the acaricide is selected to be safe for the rhino, oxpeckers, vultures, and other animals in the preserve's ecosystem.

PESTICIDE RESISTANCE

Pesticide resistance describes the decreased susceptibility of a pest population to a pesticide that was previously effective at controlling the pest. Pest species evolve pesticide resistance via natural selection: the most resistant specimens survive and pass on their acquired heritable changes traits to their offspring.

Cases of resistance have been reported in all classes of pests (*i.e.* crop diseases, weeds, rodents, *etc.*), with 'crises' in insect control occurring early-on after the introduction of pesticide use in the 20th century. The Insecticide Resistance Action Committee (IRAC) definition of insecticide resistance is 'a heritable change in the sensitivity of a pest population that is reflected in the repeated failure of a product to achieve the expected level of control when used according to the label recommendation for that pest species'.

Pesticide resistance is increasing. Farmers in the US lost 7% of their crops to pests in the 1940s; over the 1980s and 1990s, the loss was 13%, even though more pesticides were being used. Over

500 species of pests have evolved a resistance to a pesticide. Other sources estimate the number to be around 1,000 species since 1945.

Although the evolution of pesticide resistance is usually discussed as a result of pesticide use, it is important to keep in mind that pest populations can also adapt to non-chemical methods of control. For example, the northern corn rootworm (*Diabrotica barberi*) became adapted to a corn-soybean crop rotation by spending the year when field is planted to soybeans in a diapause.

As of 2014, few new weed killers are near commercialization, and none with a novel, resistance-free mode of action.

Causes

Pesticide resistance probably stems from multiple factors:

- Many pest species produce large broods. This increases the probability of mutations and ensures the rapid expansion of resistant populations.

- Pest species had been exposed to natural toxins long before agriculture began. For example, many plants produce phytotoxins to protect them from herbivores. As a result, coevolution of herbivores and their host plants required development of the physiological capability to detoxify or tolerate poisons.

- Humans often rely almost exclusively on pesticides for pest control. This increases selection pressure towards resistance. Pesticides that fail to break down quickly contribute to selection for resistant strains even after they are no longer being applied.

- In response to resistance, managers may increase pesticide quantities/frequency, which exacerbates the problem. In addition, some pesticides are toxic toward species that feed on or compete with pests. This can allow the pest population to expand, requiring more pesticides. This is sometimes referred to as *pesticide trap*, or a *pesticide treadmill*, since farmers progressively pay more for less benefit.

- Insect predators and parasites generally have smaller populations and are less likely to evolve resistance than are pesticides' primary targets, such as mosquitoes and those that feed on plants. Weakening them allows the pests to flourish. Alternatively, resistant predators can be bred in laboratories.

- Pests with limited diets are more likely to evolve resistance, because they are exposed to higher pesticide concentrations and has less opportunity to breed with unexposed populations.

- Pests with shorter generation times develop resistance more quickly than others.

Examples:

Resistance has evolved in multiple species: resistance to insecticides was first documented by A. L. Melander in 1914 when scale insects demonstrated resistance to an inorganic insecticide. Between 1914 and 1946, 11 additional cases were recorded. The development of organic insecticides, such as DDT, gave hope that insecticide resistance was a dead issue. However, by 1947 housefly resistance to DDT had evolved. With the introduction of every new insecticide class – cyclodienes, carbamates, formamidines, organophosphates, pyrethroids, even *Bacillus thuringiensis* – cases of resistance surfaced within two to 20 years.

- Studies in America have shown that fruit flies that infest orange groves were becoming resistant to malathion.

- In Hawaii, Japan and Tennessee, the diamondback moth evolved a resistance to *Bacillus thuringiensis* about three years after it began to be used heavily.

- In England, rats in certain areas have evolved resistance that allows them to consume up to five times as much rat poison as normal rats without dying.

- DDT is no longer effective in preventing malaria in some places.

- In the southern United States, *Amaranthus palmeri*, which interferes with cotton production, has evolved resistance to the herbicide glyphosate.

- The Colorado potato beetle has evolved resistance to 52 different compounds belonging to all major insecticide classes. Resistance levels vary across populations and between beetle life stages, but in some cases can be very high (up to 2,000-fold).

- The cabbage looper is an agricultural pest that is becoming increasingly problematic due to its increasing resistance to *Bacillus thuringiensis,* as demonstrated in Canadian greenhouses. Further research found a genetic component to Bt resistance.

Multiple and Cross-resistance

- Multiple-resistance pests are resistant to more than one class of pesticide. This can happen when pesticides are used in sequence, with a new class replacing one to which pests display resistance with another.

- Cross-resistance, a related phenomenon, occurs when the genetic mutation that made the pest resistant to one pesticide also makes it resistant to others, often those with a similar mechanism of action.

Adaptation

Pests becomes resistant by evolving physiological changes that protect them from the chemical.

One protection mechanism is to increase the number of copies of a gene, allowing the organism to produce more of a protective enzyme that breaks the pesticide into less toxic chemicals. Such enzymes include esterases, glutathione transferases, and mixed microsomal oxidases.

Alternatively, the number and sensitivity of biochemical receptors that bind to the pesticide may be reduced.

Behavioral resistance has been described for some chemicals. For example, some *Anopheles* mosquitoes evolved a preference for resting outside that kept them away from pesticide sprayed on interior walls.

Resistance may involve rapid excretion of toxins, secretion of them within the body away from vulnerable tissues and decreased penetration through the body wall.

Mutation in only a single gene can lead to the evolution of a resistant organism. In other cases, multiple genes are involved. Resistant genes are usually autosomal. This means that they are located on autosomes (as opposed to allosomes, also known as sex chromosomes). As a result, resistance is inherited similarly in males and females. Also, resistance is usually inherited as an incompletely dominant trait. When a resistant individual mates with a susceptible individual, their progeny generally has a level of resistance intermediate between the parents.

Adaptation to pesticides comes with an evolutionary cost, usually decreasing relative fitness of organisms in the absence of pesticides. Resistant individuals often have reduced reproductive output, life expectancy, mobility, etc. Non-resistant individuals grow in frequency in the absence of pesticides, offering one way to combat resistance.

Blowfly maggots produce an enzyme that confers resistance to organochloride insecticides. Scientists have researched ways to use this enzyme to break down pesticides in the environment, which would detoxify them and prevent harmful environmental effects. A similar enzyme produced by soil bacteria that also breaks down organochlorides works faster and remains stable in a variety of conditions.

Management

Integrated pest management (IPM) approach provides a balanced approach to minimizing resistance.

Resistance can be managed by reducing use of a pesticide. This allows non-resistant organisms to out-compete resistant strains. They can later be killed by returning to use of the pesticide.

A complementary approach is to site untreated refuges near treated croplands where susceptible pests can survive.

When pesticides are the sole or predominant method of pest control, resistance is commonly managed through pesticide rotation. This involves switching among pesticide classes with different modes of action to delay or mitigate pest resistance. The U.S. Environmental Protection Agency (EPA) designates different classes of fungicides, herbicides and insecticides. Manufacturers may recommend no more than a specified number of consecutive applications of a pesticide class be made before moving to a different pesticide class.

Two or more pesticides with different modes of action can be tankmixed on the farm to improve results and delay or mitigate existing pest resistance.

Status

Glyphosate

Glyphosate-resistant weeds are now present in the vast majority of soybean, cotton, and corn farms in some U.S. states. Weeds resistant to multiple herbicide modes of action are also on the rise.

Before glyphosate, most herbicides would kill a limited number of weed species, forcing farmers to continually rotate their crops and herbicides to prevent resistance. Glyphosate disrupts the ability of most plants to construct new proteins. Glyphosate-tolerant transgenic crops are not affected.

A weed family that includes waterhemp (*Amaranthus rudis*) has developed glyphosate-resistant strains. A 2008 to 2009 survey of 144 populations of waterhemp in 41 Missouri counties revealed glyphosate resistance in 69%. Weed surveys from some 500 sites throughout Iowa in 2011 and 2012 revealed glyphosate resistance in approximately 64% of waterhemp samples.

In response to the rise in glyphosate resistance, farmers turned to other herbicides—applying several in a single season. In the United States, most midwestern and southern farmers continue to use glyphosate because it still controls most weed species, applying other herbicides, known as residuals, to deal with resistance.

The use of multiple herbicides appears to have slowed the spread of glyphosate resistance. From 2005 through 2010 researchers discovered 13 different weed species that had developed resistance to glyphosate. From 2010-2014 only two more were discovered.

A 2013 Missouri survey showed that multiply-resistant weeds had spread. 43% of the sampled weed populations were resistant to two different herbicides, 6% to three and 0.5% to four. In Iowa a survey revealed dual resistance in 89% of waterhemp populations, 25% resistant to three and 10% resistant to five.

Resistance increases pesticide costs. For southern cotton, herbicide costs climbed from between $50 and $75 per hectare a few years ago to about $370 per hectare in 2014. In the South, resistance contributed to the shift that reduced cotton planting by 70% in Arkansas and 60% in Tennessee. For soybeans in Illinois, costs rose from about $25 to $160 per hectare.

Bacillus thuringiensis

During 2009 and 2010, some Iowa fields showed severe injury to corn producing Bt toxin Cry3Bb1 by western corn rootworm. During 2011, mCry3A corn also displayed insect damage, including cross-resistance between these toxins. Resistance persisted and spread in Iowa. Bt corn that targets western corn rootworm does not produce a high dose of Bt toxin, and displays less resistance than that seen in a high-dose Bt crop.

Products such as Capture LFR (containing the pyrethroid Bifenthrin) and SmartChoice (containing a pyrethroid and an organophosphate) have been increasingly used to complement Bt crops that farmers find alone to be unable to prevent insect-driven injury. Multiple studies have found the practice to be either ineffective or to accelerate the development of resistant strains.

PESTICIDE RESIDUE

Pesticide residue refers to the pesticides that may remain on or in food after they are applied to food crops. The maximum allowable levels of these residues in foods are often stipulated by regulatory bodies in many countries. Regulations such as pre-harvest intervals also often prevent harvest of crop or livestock products if recently treated in order to allow residue concentrations to decrease over time to safe levels before harvest. Exposure of the general population to these residues most commonly occurs through consumption of treated food sources, or being in close contact to areas treated with pesticides such as farms or lawns.

Many of these chemical residues, especially derivatives of chlorinated pesticides, exhibit bioaccumulation which could build up to harmful levels in the body as well as in the environment. Persistent chemicals can be magnified through the food chain and have been detected in products ranging from meat, poultry, and fish, to vegetable oils, nuts, and various fruits and vegetables.

A pesticide is a substance or a mixture of substances used for killing pests: organisms dangerous to cultivated plants or to animals. The term applies to various pesticides such as insecticide, fungicide, herbicide and nematocide. Applications of pesticides to crops and animals may leave residues in or on food when it is consumed, and those specified derivatives are considered to be of toxicological significance.

From post-World War II era, chemical pesticides have become the most important form of pest control. There are two categories of pesticides, first-generation pesticides and second-generation pesticide. The first-generation pesticides, which were used prior to 1940, consisted of compounds such as arsenic, mercury, and lead. These were soon abandoned because they were highly toxic and ineffective. The second-generation pesticides were composed of synthetic organic compounds. The growth in these pesticides accelerated in late 1940s after Paul Müller discovered DDT in 1939. The effects of pesticides such as aldrin, dieldrin, endrin, chlordane, parathion, captan and 2,4-D were also found at this time. Those pesticides were widely used due to its effective pest control. However, in 1946, people started to resist to the widespread use of pesticides, especially DDT since it harms non-target plants and animals. People became aware of problems with residues and its potential health risks. In the 1960s, Rachel Carson wrote *Silent Spring* to illustrate a risk of DDT and how it is threatening biodiversity.

Health Impacts

Many pesticides achieve their intended use of killing pests by disrupting the nervous system. Due to similarities in brain biochemistry among many different organisms, there is much speculation that these chemicals can have a negative impact on humans as well. There are epidemiological studies that show positive correlations between exposure to pesticides through occupational hazard, which tends to be significantly higher than that ingested by the general population through food, and the occurrence of certain cancers. Although most of the general population may not exposed to large portion of pesticides, many of the pesticide residues that are attached tend to be lipophilic and can bio-accumulate in the body.

According to the American Cancer Society there is no evidence that pesticide residues increase the risk of people getting cancer. They advise washing fruit and vegetables before eating to remove both pesticide residue and other undesirable contaminants.

In China, a number of incidents have occurred where state limits were exceeded by large amounts or where the wrong pesticide was used. In August 1994, a serious incident of pesticide poisoning of sweet potato crops occurred in Shandong province, China. Because local farmers were not fully educated in the use of insecticides, they used the highly-toxic pesticide named parathion instead of trichlorphon. It resulted in over 300 cases of poisoning and 3 deaths. Also, there was a case where a large number of students were poisoned and 23 of them were hospitalized because of vegetables that contained excessive pesticide residues.

Child Neurodevelopment

Children are thought to be especially vulnerable to exposure to pesticide residues, especially if exposure occurs at critical windows of development. Infants and children consume higher amounts of food and water relative to their body-weight have higher surface area (i.e. skin surface) relative to their volume, and have a more permeable blood-brain barrier, and engage in behaviors like crawling and putting objects in their mouths, all of which can contribute to increased risks from exposure to pesticide residues through food or environmental routes. Neurotoxins and other chemicals that originate from pesticides pose the biggest threat to the developing human brain and nervous system. Presence of pesticide metabolites in urine samples have been implicated in disorders such as attention deficit hyperactivity disorder (ADHD), autism, behavioral and emotional problems, and delays in development. There is a lack of evidence of a direct cause-and-effect relationship between long-term, low-dose exposure to pesticide residues and neurological disease, partly because manufacturers are not always legally required to examine potential long-term threats.

PESTICIDES FORMULATION

A pesticide formulation is a mixture of chemicals which effectively controls a pest. Formulating a pesticide involves processing it to improve its storage, handling, safety, application, or effectiveness.

Pesticide chemicals in their "raw" or unformulated state are not usually suitable for pest control. These concentrated chemicals and active ingredients may not mix well with water, may be chemically unstable, and may be difficult to handle and transport. For these reasons, manufacturers add inert substances, such as clays and solvents, to improve application effectiveness, safety, handling, and storage. Inert ingredients do not possess pesticidal activity and are added to serve as a carrier for the active ingredient. Manufacturers will list the percentage of inert ingredients in the formulation or designate them as "other ingredients" on their labels. There are several inert substances, such as petroleum distillates and xylene, which will have a specific statement identifying their presence in the formulation. The mixture of active and inert ingredients is called a pesticide formulation. This formulation may consist of:

- The pesticide active ingredient that controls the target pest;

- The carrier, such as an organic solvent or mineral clay;

- Adjuvants, such as stickers and spreaders;

- Other ingredients, such as stabilizers, safeners, dyes, and chemicals that improve or enhance pesticidal activity.

Herbicide

FOR THE CONTROL OF CERTAIN BROADLEAF WEEDS IN CORN (FIELD AND POP), SORGHUM (GRAIN AND FORAGE), WHEAT, BARLEY, OATS, RYE AND TRITICALE, SEEDLING ALFALFA, FLAX, GARLIC, MINT, ONIONS (DRY BULB), GRASSES GROWN FOR SOD PRODUCTION, NON-RESIDENTIAL TURFGRASS, AND NON-CROPLAND/INDUSTRIAL SITES.

ACTIVE INGREDIENT:
Octanoic acid ester of bromoxynil* (3,5-dibromo-4-hydroxybenzonitrile)..28%

Heptanoic acid ester of bromoxynil (3,5-dibromo-4-hydroxybenzonitrile)..27%

INERT INGREDIENTS: 45%
Contains xylene range/petroleum distillates.

*Equivalent to not less than 4.0 pounds of bromoxynil per gallon.

Some inert ingredients are specifically identified in the label ingredient statement.

Usually you need to mix a formulated product with water or oil for final application. Most baits, granules, gels, and dusts, however, are ready for use without additional dilution. Manufacturers package many specialized pesticides, such as products for households, in ready-to-use formulations.

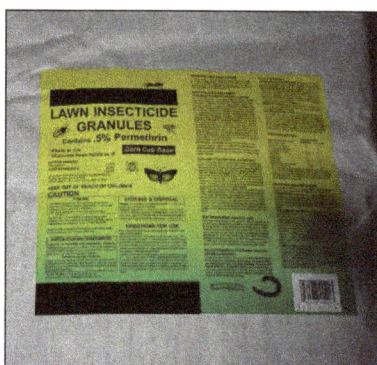

Granular, ready-to-use formulation.

A single active ingredient often is sold in several kinds of formulations. Abbreviations are frequently used to describe the formulation (e.g., WP for wettable powders); how the pesticide is used (e.g., TC for termiticide concentrate); or the characteristics of the formulation (e.g., ULV for an ultra-low-volume formulation). The amount of active ingredient (a.i.) and the kind of formulation are listed on the product label. For example, an 80% SP contains 80 percent by weight of active ingredient and is a "soluble powder." If it is in a 10-pound bag, it contains 8 pounds of a.i. and 2 pounds of inert ingredient. Liquid formulations indicate the amount of a.i. in pounds per gallon. For example, 1E means 1 pound, and 4E means 4 pounds of the a.i. per gallon in an emulsifiable concentrate formulation.

Emulsifiable concentrate formulation (4 pounds a.i. per gallon).

If you find that more than one formulation is available for your pest control situation, you should choose the best one for the job. Before you make the choice, ask several questions about each formulation. For example:

- Do we have the necessary application equipment?

- Can the formulation be applied appropriately under the conditions in the application area?

- Will the formulation reach the target and stay in place long enough to control the pest?

- Is the formulation likely to damage the surface to which you will apply it?

- Could we choose a less hazardous formulation that would still be as effective?

To answer these kinds of questions, you need to know something about the characteristics of different types of formulations and the general advantages and disadvantages of each type.

Liquid Formulations

Liquid formulations are generally mixed with water, but in some instances labels may permit the use of crop oil, diesel fuel, kerosene, or some other light oil as a carrier. This topic will present more detailed information about the common liquid pesticide formulations.

Emulsifiable Concentrates (EC or E)

An emulsifiable concentrate formulation usually contains a liquid active ingredient, one or more petroleum-based solvents (which give EC formulations their strong odor), and an agent—known as an emulsifier—that allows the formulation to be mixed with water to form an emulsion. Upon mixing with water, they take on a "milky" appearance.

Undiluted and diluted emulsifiable concentrate formulation.

Most ECs contain between 25% and 75% (2–8 pounds) active ingredient per gallon. ECs are among the most versatile formulations. They are used against agricultural, ornamental and turf, forestry, structural, food processing, livestock, and public health pests. They are adaptable to many types of application equipment including portable sprayers, hydraulic sprayers, low-volume ground sprayers, mist blowers, and low-volume aircraft sprayers.

Advantages of emulsifiable concentrates include:

- Relatively easy to handle, transport, and store.

- Little agitation required; will not settle out or separate when equipment is running.

- Not abrasive.

- Will not plug screens or nozzles.

- Little visible residue on treated surfaces.

Their disadvantages:

- High a.i. concentration makes it easy to overdose or underdose through mixing or calibration errors.

- Easily absorbed through skin of humans or animals.

- Solvents may cause rubber or plastic hoses, gaskets, and pump parts and surfaces to deteriorate.

- May cause pitting or discoloration of painted finishes.

- Flammable—should be used and stored away from heat or open flame.

- May be corrosive.

Solutions (S)

Some pesticide active ingredients dissolve readily in a liquid carrier, such as water or a petroleum-based solvent. When mixed with the carrier, they form a solution that does not settle out or separate. Formulations of these pesticides usually contain the active ingredient, the carrier, and one or more other ingredients.

Undiluted and diluted solution formulation.

Ready-to-use Low Concentration Solutions (RTU)

Low-concentrate RTU formulations are ready to use and require no further dilution before application. They consist of a small amount of active ingredient (often 1% or less per unit volume) dissolved in an organic solvent. They usually do not stain fabrics nor have unpleasant odors. They are especially useful for structural and institutional pests and for household use. Major disadvantages of low-concentrate formulations include limited availability and high cost per unit of active ingredient.

Ready-to-use diluted formulation.

Ultra-low Volume (ULV)

These concentrates may approach 100% active ingredient. They are designed to be used "as is" or to be diluted with only small quantities of a specified carrier. They are used at rates of no more than 1/2 gallon per acre. These special purpose formulations are used mostly in outdoor applications, such as in agricultural, forestry, ornamental, and mosquito control programs.

Advantages of ultra-low-volume formulations include:

- Relatively easy to transport and store.
- Remain in solution; little agitation required.
- Not abrasive to equipment.
- Will not plug screens and nozzles.
- Leave little visible residue on treated surfaces

Their disadvantages:

- Difficult to keep pesticide on target—high drift hazard.
- Specialized equipment required.
- Easily absorbed through skin of humans or animals.
- Solvents may cause rubber or plastic hoses, gaskets, and pump parts and surfaces to deteriorate.
- Calibration and application must be done very carefully because of the high concentration of active ingredient.

Invert Emulsions

An invert emulsion contains a water-soluble pesticide dispersed in an oil carrier. Invert emulsions require a special kind of emulsifier that allows the pesticide to be mixed with a large volume of petroleum-based carrier, usually fuel oil. Invert emulsions aid in reducing drift. With other formulations, some spray drift results when water droplets begin to evaporate before reaching target surfaces; as a result, the droplets become very small and light. Because oil evaporates more slowly than water, invert emulsion droplets shrink less; therefore, more pesticide reaches the target. The oil helps to reduce runoff and improves rain resistance. It also serves as a sticker-spreader by improving surface coverage and absorption. Because droplets are relatively large and heavy, it is difficult to get thorough coverage on the undersides of foliage. Invert emulsions are most commonly used along rights-of-way where drift to susceptible non-target plants or sensitive areas can be a problem.

Invert emulsion formulations are useful for drift control along rights-of-way.

Flowables (F) or Liquids (L)

A flowable or liquid formulation combines many of the characteristics of emulsifiable concentrates and wettable powders. Manufacturers use these formulations when the active ingredient is a solid that does not dissolve in either water or oil. The active ingredient, impregnated on a substance such as clay, is ground to a very fine powder. The powder is then suspended in a small amount of liquid. The resulting liquid product is quite thick. Flowables and liquids share many of the features of emulsifiable concentrates, and they have similar disadvantages. They require moderate agitation to keep them in suspension and leave visible residues similar to those of wettable powders. Flowables/liquids are easy to handle and apply. Because they are liquids, they are subject to spilling and splashing. They contain solid particles, so they contribute to abrasive wear of nozzles and pumps. Flowable and liquid suspensions settle out in their containers. Always shake them thoroughly before pouring and mixing. Because flowable and liquid formulations tend to settle, manufacturers package them in containers of 5 gallons or less to make remixing easier.

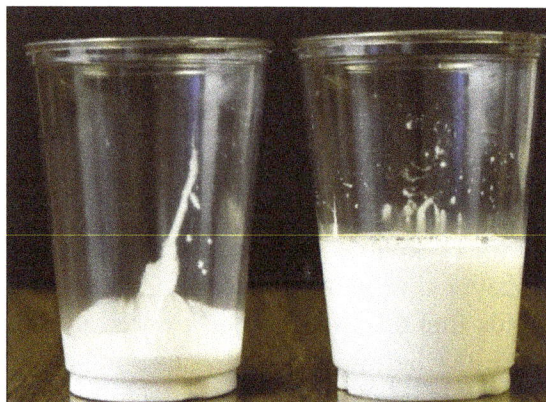

Undiluted and diluted liquid formulation.

Aerosols (A)

These formulations contain one or more active ingredients and a solvent. Most aerosols contain a low percentage of active ingredients. There are two types of aerosol formulations: the ready-to-use type commonly available in pressurized, sealed containers and those products used in electric- or gasoline-powered aerosol generators that release the formulation as a "smoke" or "fog."

Ready-to-use aerosols are usually small, self-contained units that release the pesticide when the nozzle valve is triggered. The pesticide is driven through a fine opening by an inert gas under pressure, creating fine droplets. These products are used in greenhouses, in small areas inside buildings, or in localized outdoor areas. Commercial models, which hold 5–10 pounds of pesticide, are usually refillable. Their advantages include:

- Ready to use,

- Portable,

- Easily stored,

- Convenient way to buy a small amount of a pesticide,

- Retain potency over fairly long time.

Their disadvantages:

- Practical for only very limited uses,

- Risk of inhalation injury,

- Hazardous if punctured, overheated, or used near an open flame,

- Difficult to confine to target site or pest.

Ready-to-use aerosol insecticide.

Formulations for smoke or fog generators are aerosol formulations but not under pressure. They are used in machines that break the liquid formulation into a fine mist or fog (aerosol) using a rapidly whirling disk or heated surface. These formulations are used mainly for insect control in structures such as greenhouses and warehouses and for mosquito and biting fly control outdoors.

Their advantages include:

- Easy way to fill entire enclosed space with pesticide.

Their disadvantages:

- Highly specialized use and associated equipment.

- Difficult to confine to target site or pest.

- May require respiratory protection to prevent risk of inhalation injury.

Liquid Baits

An increasing number of insecticides and rodenticides are being formulated as liquid baits. Liquid rodenticides are mixed with water and placed in bait stations designed for these products. They have two major benefits. Liquid rodenticides are effective in controlling rodents, especially rats, in areas where they cannot find water. They are also effective in areas of poor sanitation where readily available food renders traditional baits ineffective.

Liquid insecticide baits are used primarily by the structural pest control industry for controlling ants and, to a lesser extent, cockroaches. They are packaged as ready-to-use, sugar-based liquids placed inside bait stations. Liquid insecticide ant baits have a number of advantages. They are very effective against certain species of sugar-feeding ants. These ants typically accept and transfer liquid baits into the ant colonies. However, some ants will not feed on liquid baits. Liquid baits also must be replaced often.

Dry or Solid Formulations

Dry formulations can be divided into two types: ready-to-use and concentrates that must be mixed with water to be applied as a spray. This section will present more detailed information about the common dry or solid pesticide formulations.

Dusts (D)

Most dust formulations are ready to use and contain a low percentage of active ingredients (usually 10% or less by weight), plus a very fine, dry inert carrier made from talc, chalk, clay, nut hulls, or volcanic ash. The size of individual dust particles varies.

A few dust formulations are concentrates and contain a high percentage of active ingredients. These concentrates are mixed with dry inert carriers before applying.

Dusts are always used dry and can easily drift to non-target sites. They are widely used as seed treatments and sometimes for agricultural applications. In structures, dust formulations are used in cracks and crevices and for spot treatments to control insects such as cockroaches. Insects ingest poisonous dusts during grooming or absorb the dusts through their outer body covering. Dusts also are used to control lice, fleas, and other parasites on pets and livestock. Advantages of dust formulations include:

- Most are ready to use, with no mixing.

- Effective where moisture from a spray might cause damage.

- Require simple equipment.

- Effective in hard-to-reach indoor areas.

Their disadvantages:

- Easily drift off target during application.

- Residue easily moved off target by air movement or water.

- May irritate eyes, nose, throat, and skin.

- Will not stick to surfaces as well as liquids.

- Dampness can cause clogging and lumping.

- Difficult to get an even distribution of particles on surfaces.

Special dusts, known as tracking powders, are used for monitoring and controlling rodents and insects. For rodent control, the tracking powder consists of finely ground dust combined with a stomach poison. Rodents walk through the dust, pick it up on their feet and fur, and ingest it when they clean themselves. Tracking powders are useful when bait acceptance is poor because of an abundant, readily available food supply. Nontoxic powders, such as talc or flour, often are used to monitor and track the activity of rodents in buildings.

Baits (B)

A bait formulation is an active ingredient mixed with food or another attractive substance. The bait either attracts the pests or is placed where the pests will find it. Federal regulations require that certain rodenticide baits must be contained in tamper-resistant bait stations. Pests are killed by eating the bait that contains the pesticide. The amount of active ingredient in most bait formulations is quite low, usually less than 5%.

Tamper-resistant bait station.

Baits are used inside buildings to control ants, roaches, flies, other insects, and rodents. Outdoors they sometimes are used to control snails, slugs, and insects such as ants and termites. Their main use is for control of vertebrate pests such as rodents, other mammals, and birds. Advantages of baits include:

- Ready to use.

- Entire area need not be covered because pest goes to bait.

- Control pests that move in and out of an area.

Their disadvantages:

- Can be attractive to children and pets.

- May kill domestic animals and non-target wildlife outdoors.

- Pest may prefer the crop or other food to the bait.

- Dead vertebrate pests may cause odor problem.

- Other animals may be poisoned as a result of feeding on the poisoned pests.

- If baits are not removed when the pesticide becomes ineffective, they may serve as a food supply for the target pest or other pests.

- Laws require that outdoor, above-ground placement of certain rodenticide bait products be contained in tamper-resistant bait stations.

Pastes and gels are mainly used in the pest control industry for ants and cockroaches. Insecticides formulated as pastes and gels are now the primary formulations used in cockroach control. They are designed to be injected or placed as either a bead or dot inside small cracks and crevices of building elements where insects tend to hide or travel. Two basic types of tools are used to apply pastes and gels: syringes and bait guns. The applicator forces the bait out of the tip of the device by applying pressure to a plunger or trigger.

Granules (G)

Granular formulations are similar to dust formulations except granular particles are larger and heavier. The coarse particles are made from materials such as clay, corncobs, or walnut shells. The active ingredient either coats the outside of the granules or is absorbed into them. The amount of active ingredient is relatively low, usually ranging from less than 1 to 15 percent by weight.

Granular pesticide formulation.

Granular formulations typically contain a low a.i. content.

Granular pesticides are most often used to apply chemicals to the soil to control weeds, fire ants, nematodes, and insects living in the soil or for absorption into plants through the roots. Granular formulations are sometimes applied by airplane or helicopter to minimize drift or to penetrate dense vegetation. Once applied, granules release the active ingredient slowly. Some granules require soil moisture to release the active ingredient. Granular formulations also are used to control larval mosquitoes and other aquatic pests. Granules are used in agricultural, structural, ornamental, turf, aquatic, right-of-way, and public health (biting insect) pest control operations.

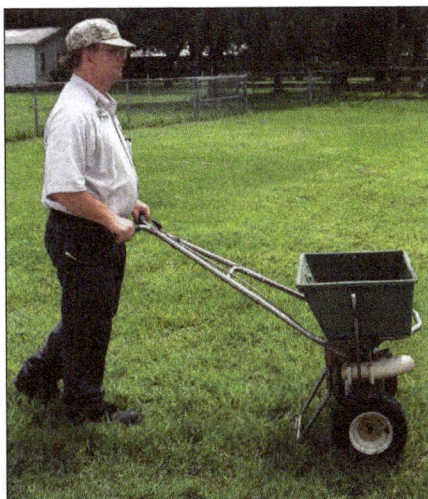

Granular formulations are often applied to the soil.

Advantages of granular formulations include:

- Ready to use, no mixing.

- Drift hazard is low, and particles settle quickly.

- Little hazard to applicator; no spray, little dust.

- Weight carries the formulation through foliage to soil or water target.

- Simple application equipment needed, such as seeders or fertilizer spreaders.

- May break down more slowly than WPs or ECs because of a slow-release coating.

Their disadvantages:

- Often difficult to calibrate equipment and apply uniformly.

- Will not stick to foliage or other uneven surfaces.

- May need to be incorporated into soil or planting medium.

- May need moisture to activate pesticide.

- May be hazardous to non-target species, especially waterfowl and other birds that mistakenly feed on the seed-like granules.

- May not be effective under drought conditions because the active ingredient is not released in sufficient quantity to control the pest.

Pellets (P or PS)

Most pellet formulations are very similar to granular formulations; the terms often are used interchangeably. In a pellet formulation, however, all the particles are the same weight and shape. The uniformity of the particles allows use with precision application equipment. A few fumigants are formulated as pellets; some may be referred to as tablets. However, these are clearly labeled as fumigants. Do not confuse them with non-fumigant pellets.

Pellet formulation.   Fumigant formulated as tablets (pellets).

Wettable Powders (WP or W)

Wettable powders are dry, finely ground formulations that look like dusts. They usually must be mixed with water for application as a spray. A few products, however, may be applied either as a dust or as a wettable powder; the choice is left to the applicator. Wettable powders contain 5%–95% active ingredient by weight, usually 50% or more. The particles do not dissolve in water. They settle out quickly unless constantly agitated to keep them suspended. Wettable powders are one of the most widely used pesticide formulations. They can be used for most pest problems and in most types of spray equipment where agitation is possible. Wettable powders have excellent residual activity. Because of their physical properties, most of the pesticide remains on the surface of treated porous materials such as concrete, plaster, and untreated wood. In such cases, only the water penetrates the material.

Undiluted and diluted wettable powder formulation.

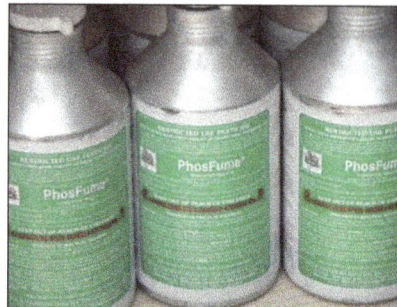

Advantages of wettable powders include:

- Easy to store, transport, and handle.

- Less likely than ECs and other petroleum-based pesticides to cause unwanted harm to treated plants, animals, and surfaces.

- Easily measured and mixed.

- Less skin and eye absorption than ECs and other liquid formulations.

Their disadvantages:

- Inhalation hazard to applicator while measuring and mixing the concentrated powder.

- Require good and constant agitation (usually mechanical) in the spray tank or will quickly settle out if the agitator is turned off.

- Abrasive to many pumps and nozzles, causing them to wear out quickly.

- Difficult to mix in very hard, alkaline water.

- Often clog nozzles and screens.

- Residues may be visible on treated surfaces.

Soluble Powders (SP or WSP)

Soluble powder formulations look like wettable powders. However, when mixed with water, soluble powders dissolve readily and form a true solution. After they are mixed thoroughly, no additional agitation is necessary. The amount of active ingredient in soluble powders ranges from 15% to 95% by weight; it usually is more than 50%. Soluble powders have all the advantages of wettable powders and none of the disadvantages except the inhalation hazard during mixing. Few pesticides are available in this formulation because few active ingredients are readily soluble in water.

Water-dispersible Granules (WDG)

Water-dispersible granules, also known as dry flowables, are like wettable powders except instead of being dustlike, they are formulated as small, easily measured granules. Water-dispersible granules must be mixed with water to be applied. Once in water, the granules break apart into fine particles similar to wettable powders. The formulation requires constant agitation to keep them suspended in water. The percentage of active ingredient is high, often as much as 90 percent by weight. Water-dispersible granules share many of the same advantages and disadvantages of wettable powders except:

- They are more easily measured and mixed.

- Because of low dust, they cause less inhalation hazard to the applicator during handling.

Undiluted and diluted dry flowable formulation.

Other Formulations

Other formulations include chemicals that cannot be clearly classified as liquid or as dry/solid pesticide formulations.

Microencapsulated Materials (M or ME)

Manufacturers cover liquid or dry pesticide particles in a plastic coating to produce a microencapsulated formulation. Microencapsulated pesticides are mixed with water and sprayed in the same manner as other sprayable formulations. After spraying, the plastic coating breaks down and slowly releases the active ingredient. Microencapsulated materials have several advantages:

- Highly toxic materials are safer for applicators to mix and apply.

- Delayed or slow release of the active ingredient prolongs its effectiveness, allowing for fewer and less precisely timed applications.

- The pesticide volatilizes more slowly; less is lost from the application site.

In residential, industrial, and institutional applications, microencapsulated formulations offer several advantages. These include reduced odor, the release of small quantities of pesticide over a long time, and greater safety. Microencapsulated materials offer fewer hazards to the skin than ordinary formulations. Microencapsulated materials, however, pose a special hazard to bees. Foraging bees may carry microencapsulated materials back to their hives because they are about the same size as pollen grains. As the capsules break down, they release the pesticide, poisoning the adults and brood.

Breakdown of the microencapsulated materials to release the pesticide sometimes depends on weather conditions. Under certain conditions, the microencapsulated materials may break down more slowly than expected. This could leave higher residues of pesticide active ingredient in treated areas beyond normal restricted-entry or harvest intervals with the potential to injure fieldworkers. For this reason, regulations require long restricted-entry intervals for some microencapsulated formulations.

Water-soluble Packets (WSB or WSP)

Water-soluble packets reduce the mixing and handling hazards of some highly toxic pesticides. Manufacturers package precise amounts of wettable powder or soluble powder formulations in a special type of plastic bag. When you drop these bags into a filled spray tank, they dissolve and release their contents to mix with the water. There are no risks of inhaling or contacting the undiluted pesticide as long as you do not open the packets. Once mixed with water, however, pesticides packaged in water-soluble packets are no safer than other diluted pesticides.

Water-soluble packet dissolves in the tank.

Attractants

Attractants include pheromones, sugar and protein syrups, yeasts, and rotting meat. Pest managers use these attractants in various types of traps. Attractants also can be combined with pesticides and sprayed onto foliage or other items in the treatment area.

Insect cone trap containing pheromone.

Impregnated Products

Manufacturers impregnate (saturate) pet collars, livestock ear tags, adhesive tapes, plastic pest strips, and other products with pesticides. These pesticides evaporate over time, and the vapors provide control of nearby pests. Some paints and wood finishes have pesticides incorporated into them to kill insects or retard fungal growth. Fertilizers also may be impregnated with pesticides.

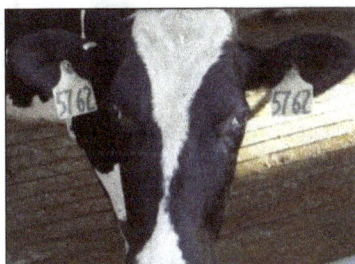

Cattle ear tag impregnated with insecticide.

Repellents

Various types of insect repellents are available in aerosol and lotion formulations. People apply these to their skin or clothing or to plant foliage to repel biting and nuisance insects. You can mix other types of repellents with water and spray them onto ornamental plants and agricultural crops to prevent damage from deer, dogs, and other animals.

Animal Systemic

Systemic pesticides protect animals against fleas and other external blood-feeding insects as well as against worms and other internal parasites. A systemic animal pesticide is one that is absorbed and moves within the animal. These pesticides enter the animal's tissues after being applied orally or externally. Oral applications include food additives and premeasured capsules and liquids. External applications involve pour-on liquids, liquid sprays, and dusts. Most animal systemics are used under the supervision of veterinarians.

Pesticide or Fertilizer Combinations

Pest managers frequently use insecticides, fungicides, and herbicides in combination with fertilizers. This provides a convenient way of controlling pests while fertilizing crops or lawns. Homeowners commonly use these combinations, although the unit cost of pesticide in these formulations is usually high. Dealers or growers often custom-mix pesticides with fertilizers to meet specific crop requirements.

Fumigants

Fumigants are pesticides that form gases or vapors toxic to plants, animals, and microorganisms. Some active ingredients are formulated, packaged, and released as gases; others are liquids when packaged under high pressure and change to gases when they are released. Other active ingredients are volatile liquids when enclosed in an ordinary container and, therefore, are not formulated under pressure. Others are solids that release gases when applied under conditions of high humidity or in the presence of water vapor. Fumigants are used for structural pest control, in food and grain storage facilities, and in regulatory pest control at ports of entry and at state and national borders. In agricultural pest control, fumigants are used in soil, greenhouses, granaries, and grain bins.

Gaseous fumigant stored under pressure.

Structural fumigation of a dormitory on UF campus.

Soil fumigant application.

Advantages of fumigants:

- Toxic to a wide range of pests.

- Can penetrate cracks, crevices, wood, and tightly packed areas such as soil or stored grains.

- Single treatment usually kills most pests in treated area.

Disadvantages of fumigants:

- The target site must be enclosed or covered to prevent the gas from escaping.

- Nonspecific in that they are highly toxic to humans and all other living organisms.

- Require the use of specialized protective equipment, including respirators specifically approved for use with fumigants.

- Require the use of specialized application equipment.

Adjuvants

Adjuvants are substances used with a pesticide to enhance performance. By themselves, they do not possess pesticidal activity. Adjuvants may be added to the product at the time of formulation or by the applicator to the spray mix just prior to treatment. Adjuvants include surfactants, compatibility agents, antifoaming agents and spray colorants (dyes), and drift control agents.

Much of the confusion surrounding adjuvants can be attributed to the lack of understanding of adjuvant terminology. For example, many people use the terms adjuvant and surfactant interchangeably. These terms can refer to the same product because all surfactants are adjuvants. However, not all adjuvants are surfactants.

Care should be taken when selecting an adjuvant. Pesticide performance can differ depending on what type of adjuvant is used. The pesticide label will state if specific surfactants are required and the amount (%) of active ingredient it must contain.

VARIOUS METHODS IN PESTICIDE FORMULATION ANALYSIS

Titrimetry

Titration, also known as titrimetry, is a common laboratory method of quantitative chemical analysis that is used to determine the unknown concentration of an identified analyte. Because volume measurements play a key role in titration, it is also known as volumetric analysis. A reagent, called the titrant or titrator is prepared as a standard solution. A known concentration and volume of titrant reacts with a solution of analyte or titrand to determine concentration.

Small volumes of the titrant are then added to the titrand and indicator until the indicator changes, reflecting arrival at the endpoint of the titration. Depending on the endpoint desired, single drops or less than a single drop of the titrant can make the difference between a permanent and temporary change in the indicator. When the endpoint of the reaction is reached, the volume of reactant consumed is measured and used to calculate the concentration of analyte.

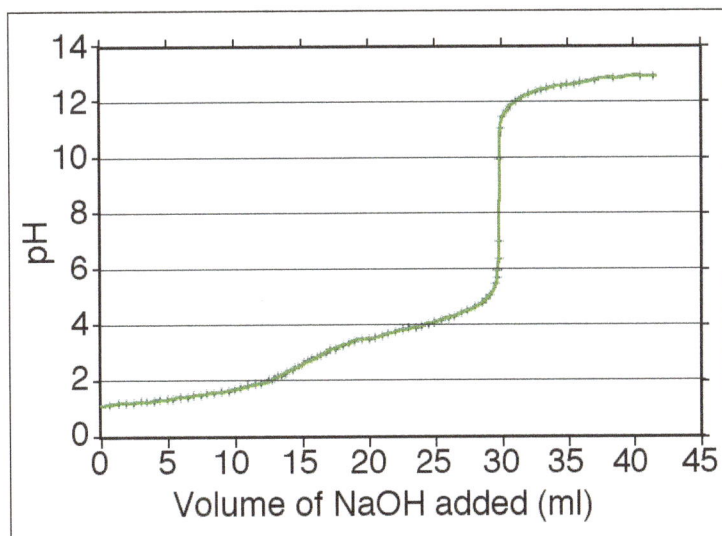

Titration Curve.

It is appropriate to know the following terminology used in the titrimetric analysis:

- Titrant: The solution containing the active agent with which a titration is made.

- Titrand: The solution containing the active agent which is titrated.

- Equivalence-point: The point in a titration at which the amount of titrant added is chemically equivalent to the amount of substance titrated. (Stoichiometric-point and Theoretical end-point are synonymous with Equivalence-point.)

- End-point: The point in a titration at which some property of the solution (as, for example, the colour imparted by an indicator) shows a pronounced change, corresponding more or less closely to the equivalencepoint. The end-point may be represented by the intersection of two lines or curves in the graphical method of end-point determination.

Standard Substance

- Primary standard: A substance of high purity which, by stoichiometric reaction, is used to establish the reacting strength of a titrant, or which itself can be used to prepare a titrant solution of accurately known concentration.

- Secondary standard: A substance used for standardizations, whose content of the active agent has been found by comparison against a primary standard.

- Standardisation: The process of finding the concentration or an active agent in a solution, or the reacting strength of a solution in terms of some substance, usually by titration of a known amount of the substance which is pure or has a known reaction value.

- Indirect Titration: A titration (acid—base or other type) in which the entity being determined does not react directly with the titrant, but indirectly via the intermediacy of a stoichiometric reaction with another titratable entity.

- Standard solution: A solution having an accurately known concentration of the active substance, or an accurately known titre.

- Primary standard solution: A standard solution prepared from a primary standard substance whose concentration is known from the weight of that substance in a known volume (or weight) of the solution.

- Secondary standard solution: A solution whose concentration or titre has been obtained by standardization, or which has been prepared from a known weight of a secondary standard substance.

- Titration error: The difference in the amount of titrant, or the corresponding difference in the amount of substance being titrated, represented by the expression:

  (End-point value — Equivalence-point value)

Types of Titrations

- Acid-base: A titration involving the transfer of protons (Bronsted—Lowry) or electron-pairs (Lewis) from one of the reacting species to the other in solution.

- Non-aqueous: A titration (acid—base or other type) in which the solvent medium is one other than water and in which the concentration of the latter is minimal

- Compleximetric: A titration involving the formation of a soluble complex between a metal ion and a complexing agent.

- Iodimetric: Titration with, or of, iodine Some authors restrict iodimetry to titration with a standard solution of iodine, and iodometry to titration of iodine; such restrictions are not recommended.

- Oxidation-reduction (Redox): A titration involving the transfer of one or more electrons from a donor ion or molecule (the reductant) to an acceptor (the oxidant).

- Precipitation: A titration in which the entity being titrated is precipitated from solution by reaction with the titrant.

Acid-base titrations: Before starting the titration a suitable pH indicator must be chosen. The equivalence point of the reaction, the point at which equivalent amounts of the reactants have reacted, will have a pH dependent on the relative strengths of the acid and base used. The pH of the equivalence point can be estimated using the following rules:

- A strong acid will react with a strong base to form a neutral (pH=7) solution.

- A strong acid will react with a weak base to form an acidic (pH7) solution.

- A weak acid will react with a strong base to form a basic (pH>7) solution.

When a weak acid reacts with a weak base, the equivalence point solution will be basic if the base is stronger and acidic if the acid is stronger. If both are of equal strength, then the equivalence pH will be neutral. However, weak acids are not often titrated against weak bases because the colour change shown with the indicator is often quick, and therefore very difficult for the observer to see the change of colour.

Acid—base titration is performed with a phenolphthalein indicator, when it is a strong acid – strong base titration, a bromthymol blue indicator in weak acid – weak base reactions, and a methyl orange indicator for strong acid – weak base reactions.

Different methods to determine the equivalence point include:

- pH indicator: This is a substance that changes color in response to a chemical change. An acid-base indicator (e.g., phenolphthalein) changes color depending on the pH. Redox indicators are also frequently used. A drop of indicator solution is added to the titration at the start; when the color changes the endpoint has been reached, this is an approximation of the equivalence point. Some of the pH indictors are listed below.

Indicator	Color on acidic side	Range of color change	Color on basic side
Bromophenol Blue	Yellow	3.0–4.6	Blue
Methyl Orange	Red	3.1–4.4	Yellow
Congo red	Blue	3.0 - 5.0	Red
Bromocresol Green	Yellow	3.8 – 5.4	Blue
Methyl Red	Red	4.4–6.3	Yellow

Litmus	Red	5.0–8.0	Blue
Bromothymol Blue	Yellow	6.0–7.6	Blue
Phenolphthalein	Colorless	8.3–10.0	Pink
Alizarin Yellow	Yellow	10.1–12.0	Red

- Potentiometer: A potentiometer can also be used. This is an instrument which measures the electrode potential of the solution. These are used for titrations based on a redox reaction; the potential of the working electrode will suddenly change as the equivalence point is reached.

- pH meter: This is a potentiometer which uses an electrode whose potential depends on the concentration of H_3O^+ present in the solution. This is an example of an ion selective electrode. This method allows the pH of the solution to be measured throughout the titration. At the equivalence point there will be a sudden change in the measured pH. It can be more accurate than the indicator method, and is very easily automated.

- Conductance: The conductivity of a solution depends on the ions that are present in it. During many titrations, the conductivity changes significantly. (For instance, during an acid-base titration, the H_3O^+ and OHions react to form neutral H_2O. This changes the conductivity of the solution.) The total conductance of the solution depends also on the other ions present in the solution (such as counter ions). Not all ions contribute equally to the conductivity; this also depends on the mobility of each ion and on the total concentration of ions (ionic strength). Thus, predicting the change in conductivity is harder than measuring it.

- Color change: In some reactions, the solution changes colour without any added indicator. This is often seen in redox titrations, for instance, when the different oxidation states of the product and reactant produce different colours.

Complexometric titrations: A simple example in this type can be the titrations involving Silver nitrate in titrating the cyanide ion:

$$Ag^+ + 2CN^- \leftrightarrow \left[Ag(CN)_2\right]^-$$

using the indicator potassium iodide ion and ammonia solution.

$\left[Ag(NH_3)_2\right]^+ \leftrightarrow AgI + 2NH_3$ the formation silver iodide gives turbidity to the solution.

Another important complexometric titration is the titrations involving ethylenediaminetetra acetic acid (EDTA) in determining the metal ions.

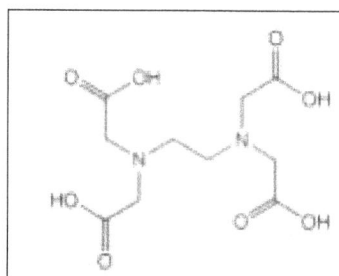

$$M^{2+} + H_4Y \rightarrow MH_2Y + 2H^+$$

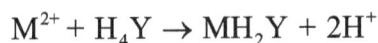

Commonly used indicators in the EDTA titrations are organic dyes such as Fast Sulphon Black, Eriochrome Black T, Eriochrome Red B or Murexide.

Iodometric Titration: Free iodine is titrated against standard reducing agent such as sodium thiosulfate solution. Usual reagents are sodium thiosulfate as titrant, starch as an indicator (it forms blue complex with free iodine molecules), and an iodine compound (iodide or iodate, depending on the desired reaction with the sample). The color change at the end point is usually blue to colorless. The principal reaction is the reduction of iodine to iodide by thiosulfate (One of the examples for redox titrations).

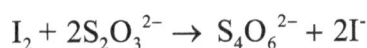

$$I_2 + 2S_2O_3^{2-} \rightarrow S_4O_6^{2-} + 2I^-$$

Redox titrations: As the name indicates both oxidation and reduction simultaneously takes place in the molecules to their respective oxidation/reduction states. The chemical reaction proceeds with transfer of electrons (simultaneous loss and gain of electrons) among the reacting ions in aqueous media.

Some Redox titrations are named after the reagent:

Permanganate Titrations

Potassium permanganate is used as an oxidizing agent. The medium is maintained acidic by the use of Dil. H_2SO_4. Potassium permanganate acts as self-indicator Used in the estimation of ferrous salts, oxalic acid, oxalates, H_2O_2 etc. Solution of the Potassium permanganate should be standardized first using standard oxalic acid or sodium oxalate.

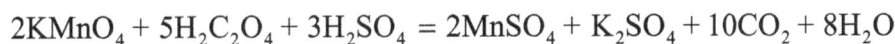

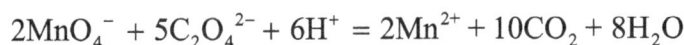

$$2KMnO_4 + 5H_2C_2O_4 + 3H_2SO_4 = 2MnSO_4 + K_2SO_4 + 10CO_2 + 8H_2O$$
$$2MnO_4^- + 5C_2O_4^{2-} + 6H^+ = 2Mn^{2+} + 10CO_2 + 8H_2O$$

Dichromate Titrations

Potassium dichromate is used as an oxidizing agent in acedic medium and sulphuric acid is used to maintain the medium acidic. Potassium dichromate solution can be directly used for titrations. This type of titration is mainly use in the estimation of Ferrous salts and Iodides. In these titrations of potassium dichromate versus ferrous salt, either an external indicator (Potassium ferricyanide) or an internal indicator (Diphenyl amine) can be used.

Iodimetric titrations also are redox titrations.

1. Precipitation reactions: The argentometric titrations involving Silver nitrate in determining halogens is a classical example for precipitation reactions. The titrant react with the analyte forming an insoluble material and the titration continues till the very last amount of analyte is consumed.

$$Ag^+ + Cl^- = AgCl$$

The first drop of titrant in excess will react with an indicator resulting in a color change and announcing the termination of the titration.

$$2Ag^+ + CrO_4^{2-} = Ag_2CrO_4$$

This reaction should be carried out in faintly alkaline solution within the pH range of 6.5 to 9.0. In acid solution, the chromate under goes the following reaction, where in no colour change can be observed.

$$CrO_4^{2-} + 2H^+ \leftrightarrow HCrO_4^- \leftrightarrow Cr_2O_7^{2-} + H_2O$$

Nonaqueous titration is the titration of substances dissolved in non-aqueous solvents. It is suitable for the titration of very week acids and very weak bases.

E.g. The titration of organic bases with perchloric acid in anhydrous acetic acid. If a very strong acid such as perchloric acid is dissolved in acetic acid, the latter can function as a base and combine with protons donated by the perchloric acid to form protonated acetic acid.

Since the $CH_3COOH_2^+$ ion readily donates its proton to a base, the titration of a base is accomplished. The reactions involved can be expressed as furnished.

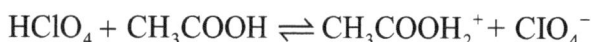

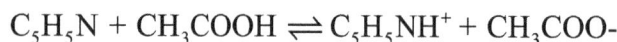

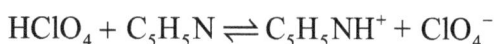

$$HClO_4 \rightleftharpoons H^+ + ClO4^-$$
$$CH_3COOH + H^+ \rightleftharpoons CH_3COOH2^+ \text{ (onium ion)}$$
$$HClO_4 + CH_3COOH \rightleftharpoons CH_3COOH_2^+ + CIO_4^-$$

$$C_5H_5N + CH_3COOH \rightleftharpoons C_5H_5NH^+ + CH_3COO-$$
$$CH_3COOH_2^+ + CH3COO^- \rightleftharpoons 2CH_3COOH$$

$$HClO_4 + C_5H_5N \rightleftharpoons C_5H_5NH^+ + ClO_4^-$$

Popularly used indicator used in this titrations is crystal violet, which shows different colours in different medium as given under:

    Violet   Blue-green   Yellowish-green

Preparation of Standard Solutions

1. Preparation of 0.1N silver nitrate: Weigh about 17.0 grams of A.R. Silver nitrate and dissolve in 1000 ml water. This is standardized against sodium chloride using potassium chromate indicator (The Mohr titration).

Standardisation:

Weigh out accurately 0.1000 to 0.1500 grams of A.R. Sodium Chloride and dissolve in sufficient amount of water (50-100 ml). Add 1 ml of the indicator. Add the silver nitrate solution slowl from the burette, swirling the liquid constantly, until the red colour formed by the addition of each drop begins to disappear more slowly. Continue the addition dropwise until a faint but distinct change in colour occurs. This faint reddish-brown colour should persist after brisk shaking.

Preparation of potassium Chromate indicator solution:

Dissolve 5 gms of A.R. Potassium Chromate in 100 ml water.

$$\text{Normality of Silver Nitrate} = \frac{\text{Weight of NaCl} \times 1000}{\text{Volume of AgNO}_3 \times 58.46}$$

2. Preparation of 0.1N Potassium Thiocyanate: Weigh out about 10.5 gms. of A.R. Potassium Thiocyanate and dissolve it in 1 litre of water.

Standardisation:

Pipette 25 ml of the standard 0.1N silver nitrate into a 250 ml conical flask adds 5 ml of 6N-nitric acid and 1 ml of ferric indicator solution. Run in the potassium thiocyanate solution from the burette. Continue the addition until one drop of the thiocyanate solution produces a faint brown colour, which no longer disappears upon shaking.

Normality of Potassium

$$\text{Thiocyanate solution} = \frac{\text{Normality of AgNO}_3 \times 25}{\text{Volume of thiocyanate added}}$$

The ferric indicator solution consists of a cold, saturated solution of A.R. ferric ammonium sulphate in water about 40% to which a few drops of 6N – nitric acid has been added.

3. Preparation of 0.1N Sodium Thiosulphate solution: Weigh about 25.0 grams of A.R. Sodium Thiosulphate $Na_2S_2O_3$. 5 H_2O, dissolve in 1 litre of water. If the solution is to be kept for more than a few days, add 0.1 gram of Sodium Carbonate or 3 drops of Chloroform.

Standardisation:

Potassium Iodate can be used for the standardization of Sodium Thiosulphate. Weigh out accurately 0.14-0.15 gm of pure dry potassium iodate, dissolve it in 25 ml water, add 2 g of iodate-free potassium iodide and 5 ml of 2 N sulphuric acid. Titrate the liberated iodine with the thiosulphate solution when the colour of the solution has become a pale yellow, dilute to 200 ml with water, add 2 ml of starch solution and continue the titration until the colour changes from blue to colourless.

$$\text{Normality of Sodium Thiosulphate} = \frac{\text{Wt. of Pot.iodate} \times 1000}{\text{Vol. of thiosulphate} \times 35.67}$$

4. Preparation of 0.1N Iodine: Dissolve 20 g. of iodate free potassium iodide in 30-40ml water. Weigh out about 12.7 g. of iodine on a watch glass on a rough balance and transfer it into the concentrated Potassium iodide solution. Shake until all the iodine has dissolved. Dilute it by water so that the total volume becomes 1 litre.

Standardisation:

Transfer 25 ml of the iodine solution to a 250 ml conical flask, dilute to 100ml and add standard sodium thiosulphate solution from a burette until the soln. has a pale yellow colour. Add 2 ml of starch solution and continue the addition of the thiosulphate solution slowly until the solution is just colourless.

$$\text{Normality of Iodine} = \frac{\text{Normality of } Na_2S_2O_3 \ \times \ \text{vol. of } Na_2S_2O_3}{25}$$

Preparation of Starch solution:

Make a paste of 1.0 g. of soluble starch with a little water, and pour the paste with constant stirring, into 100ml of boiling water and boil for 1 minute. Allow the solution to cool and add 2-3 g. of potassium iodide.

5. Preparation of 0.1N Sodium Hydroxide: Dissolve about 4.0 g. (20 pellets) of Sodium Hydroxide in 1 litre of water.

Standardisation:

This is standardized using Potassium Hydrogen Phthalate. Weight out accurately about 0.6-0.7 g. of A.R. Potassium Hydrogen Phthalate into a 250 ml conical flask, add 75 ml of water and shake gently until the solid has dissolved. Titrate this against Sodium hydroxide, using Phenolphthalein as indicator. The end point is the appearance of pink colour.

$$\text{Normality of NaOH} = \frac{\text{Wt. of potassium Hydrogen Phthalate} \times 1000}{\text{Volume of NaOH} \times 204.22}$$

6. Preparation of 0.1N Hydrochloric acid: Measure out by means of a graduated cylinder 9 ml of pure concentrated hydrochloric acid into a 1000 ml measuring cylinder containing about 500ml water. Make up to the litre mark with water and thoroughly mix by shaking.

Standardisation:

Weigh out accurately 0.2 g. of pure sodium carbonate into a 250 ml conical flask, dissolve it in 50-75 ml water and add 2 drops of methyl orange indicator. Titrate against hydrochloric acid from the burette until the colour of the methyl orange becomes orange or a faint pink.

$$\text{Normality of HCl} = \frac{\text{Weight of } Na_2CO_3 \times 1000}{\text{Volume of HCl} \times 53}$$

Alpha Naphthyl Acetic Acid

- Type of pesticide: Plant growth regulator
- Molecular formula: $C_{12} H_{10} O_2$
- Molecular weight: 186.2

- Structure:

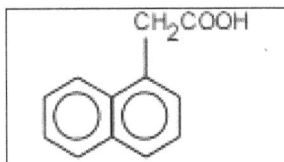

- Registered products:

  ○ Technical – 98%

  ○ S.L. – 4.5%

- Principle: Alcohol present in the SL sample is evaporated off by heating on a hot water bath. The residue is redissolved in water and acidified to release NNA from sodium salt. NNA is extracted into solvent ether and washed with distilled water till free from mineral acid. After evaporation of ether, residue of NAA is redissolved in neutral alcohol and titrated against standard NaOH.

- Reactions:

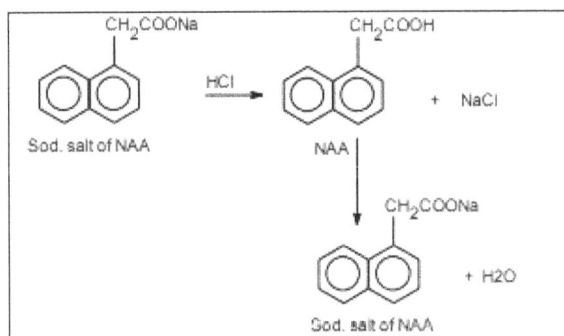

Calculations:

1 GE of NaOH = 1 gm of NAA

Therefore 1000 ml of 1(N) NaOH = 1 GE of NaOH

$$= 186.2 \ NAA$$

Therefore t ml of (n) NaOH $= \dfrac{186.2 \times t \times n}{1000}$

Therefore % AI $= \dfrac{18.62 \times t \times n}{W}$

Where,

- W = Weight in gms of the sample taken for analysis.

- T = Titration reading.

- N = Normality.

Procedure:

- Technical grade sample: 3 gms of sample, weighed accurately is dissolved in 40 ml of neutralized methanol and titrated against 0.5 (N) standard NaOH. Solution using phenolphthalein as indicator.

- S.L. Sample: 10 gm of sample is weighed accurately in a beaker and heated on a boiling water bath till solvent is evaporated off and dry residue is obtained. This residue is dissolved in minimum quantity of water and acidified using 50 ml of 5 pH buffer solution (prepared by dissolving 2.035 gms of citric acid and 2.924 gms of Na2 HPO4 in 200 ml water). The precipitate of ANA is extracted in a separating funnel, quantitatively into 50 ml of ether. Separate the water layer and reextract it twice more using 50 ml ether each time. Combine all ether layers and heat the ether layer on boiling water bath till solvent is evaporated off leaving dry residue. Dissolve the residue in 50 ml of neutralized methanol and titrate with standard 0.1(N) Na OH using phenolphthalein as indicator.

Precautions:

ANA has solubility in water to the extent of 420 mg/1. If large quantity of water is used, then there could be loss of AI.

Ether is highly volatile, inflammable and explosive. Hence, extraction should be done in cooler room. The vapour of ether should be released from time to time during extraction to prevent development of pressure and bursting of separating funnel used for extractions. Further there should be no naked flame or spark while working with ether. Never heat ether layer on hot plate or gas burner.

Carbofuran

- Type of pesticide: Systemic insecticide belonging to carbamate group

- Molecular formula: $C_{12}H_{15}O_3N$

- Molecular weight: 221.3

- Structure:

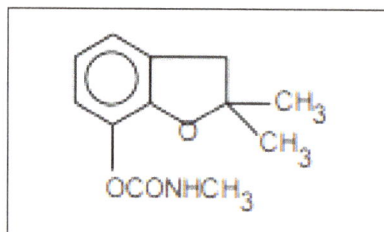

- Registered products: Tech-

    ○ 75%

    ○ 90%

- Formulations:

  ○   3% CG

  ○   50% SP for Govt. use

- Principle: One gm mole of carbofuran on alkaline hydrolysis in glycolic medium under reflux releases one gram mole of methl amine which is collected quantitatively; in 2% Boric acid solution containing of standard HCl is equivalent to hydrogen ion absorbed by methyl amine, original green colour is restord bromo cresol green indicator. The boric acid methyl amine complex is then titrated with std. HCl. When the volme. Thus hydrogen ion absorbed by methyl amine is determined which is the basis for cabculation of AI.

- Reactions:

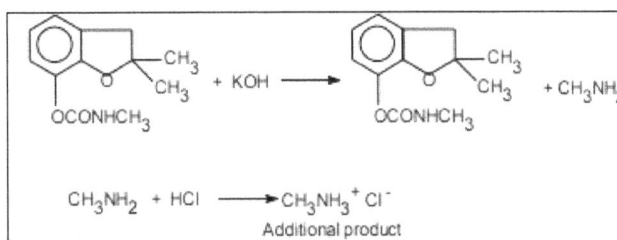

Calculations:

1 gm. eq. of HCl = 1 gm. eq. of methyl amine Or

1000 ml of 1 N HCl = 221.3 gm of carbofuran

$$T \text{ ml of N HCl} = \frac{221.3 \times t \times N}{1000} \text{ gm of AI}$$

Hence % AI = $\dfrac{22.13 \times t \times N}{W}$

Where,

- t = Titration reading in ml.

- N = Normality of HCl.

- W = Weight of sample taken.

Outline of the procedure: 0.4 to 0.6 gm of AI weighed accurately; is hydrolysed for 60 minutes under relfux and stream of nitraogen with 50 ml of 2 N glycolic KOH. The amine liberated is absorbed in 150 ml of 2% Boric acid solution containing bromocresol green as indicator and estimated by titration against standard 0.1 N HCl.

Captan

- Type of pesticide: A contact/protective fungicide with eradicant properties, belonging to phthalimide group.

- Molecular formula: $C_9H_8Cl_3NSO_2$

- Molecular weight: 301

- Structure:

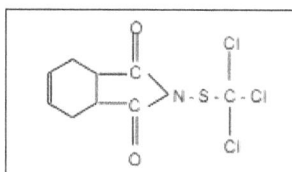

- Registered products:

    ◦ Technical = 90%

    ◦ WP = 50%

    ◦ WS = 75%

    ◦ DS = 75%

- Principle: Captan is hydrolysed with Sodium hydroxide in a medium of 1:1 mixture of methanol and acetone under reflux when 3 equivalence of ionic chloride are liberated per mole of a.i. The ionic chloride is estimated by Volhard's method. A.I content is calculated by multiplying per cent chloride (corrected free chloride), with a suitable convernsion factor.

- Reactions:

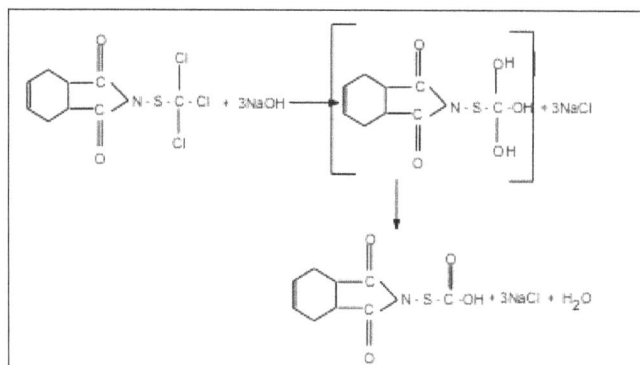

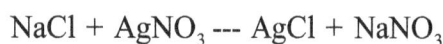

$$NaCl + AgNO_3 \text{ --- } AgCl + NaNO_3$$

Calculations:

1 gm equivalent of Silver nitrate = 1 gm equivalent of NaCl = 1/3 mole of a.i

1000ml of 1 N Silver nitrate = 100.3 gms of a.i

't' ml of N normal Silver nitrate = $\dfrac{100.3 \times t \times n}{1000}$

Hence, % A.I = $\dfrac{10.03 \times t \times n}{W}$

- Outline of the procedure: 1 gm a.i is taken in 250 ml volumetric flask and 125ml acetone is added. Shaken well to dissolve the a.i and make up to the mark with Methanol. To the 50ml aliquot, 50ml of 0.3N solution of aqueous Sodium hydroxide is added and refluxed for 1 hr.

  After cooling add 5ml Hydrogen peroxide (30%) and boil gently for 10 min. If it's still coloured, repeat the above exercise. Destroy the excess alkalinity by using 1:1 Nitric acid and phenolphthalein. Add 10ml of Nitric acid in excess, 25ml standard solution of Silver nitrate and 5ml Nitro-benzene and 1 ml Ferric alum indicator and titrated against standard Potassium thiocyanate solution to a faint brick red colour (y ml).

  Carry out a blank titration with 50ml aliquot of stock solution without hydrolysis (x ml) Calculate 't' value by deducting y ml from x ml. and use dilution factor 5.

- Precautions: Captan is extremely unstable in solution. Hence blank titration for estimating free Chloride should be done in the beginning, within half an hour of solution preparation.

Dicofol

- Type of pesticide: An acaricide with insecticidal action.

- Molecular formula: $C_{14}H_9Cl_5O$

- Molecular weight: 370.5

- Structure:

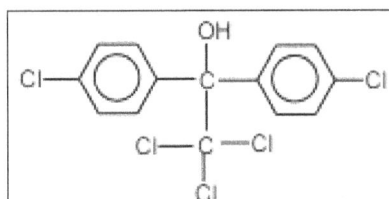

- Registered products:

  ○ Technical 82% min

  ○ EC 18.5%

- Principle: One mole of dicofol on hydrolysis with alkali under reflux yields 3 moles of ionic Chloride which is estimated by Volhard's method. Free Chloride is also estimated and the correction is applied.

- Reactions:

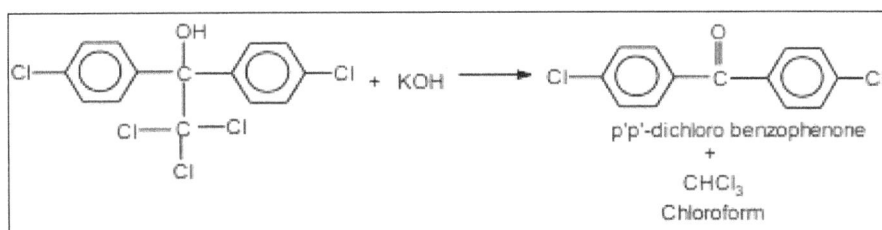

$$CHCl_3 + 4KOH \rightarrow 3KCl + HCOOK + 2H_2O$$

$$KCl + AgNO_3 \rightarrow AgCl + KNO_3$$

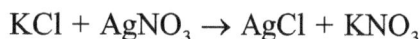

Calculations:

3 gm equivalents of Silver nitrate = 3 gm equivalents of KCl = 1 gm mole of Dicofol

1 gm equivalent of Silve nitrate = 1/3 mole of a.i = 123.5 gms of a.i

1000ml of 1 N normal Silver nitrate = 123.5 gms of a.i

't' ml of normal solution = $\dfrac{123.5 \times t \times n}{1000}$

Hence, % A.I = $\dfrac{123.5 \times t \times n \times 100}{1000 \times W} = \dfrac{12.35 \times t \times n}{W}$

- Outline of the procedure: 0.5 gms of a.i refluxed with 50 ml of 0.5 N ethanolic KOH for 90 minutes with 3 to 4 drops of ethanol as a seal. After cooling 50ml ethanol (95%) is added and swirled. The volume is made up to 300ml with distilled water, and ethanol is evoparated on water bath. Excess alkali is neutralized with 4N Nitric acid using Phenolphthalein as indicator and add 10ml Nitric acid in excess followed by 50ml of standard Silver nitrate (0.1N) is added. The excess Silver nitrate is back titrated with potassium thiocyanate.

Run a blank determination through all these steps of the procedure using all the reagents except sample.

Copper Compounds

Type of pesticide: Contact and protective fungicide.

Table: Molecular formula and Structural formula.

S. No.	Name	Structural formulae	Molecular formulae
1.	Copper oxy chloride	$3Cu(OH)_2CuCl_2$	$Cu_4H_6O_6Cl_2$
2.	Copper sulphate	$Cu SO_4 5H_2O$	$Cu H_{10}O_9S$
3.	Cuprous oxide	Cu_2O	Cu_2O

Table: Registered products.

S. No	Name	% A.I (Copper content)			
		Tech	DP	WP	Oil based
1.	Copper oxy Chloride	57	4	50	40, 50 & 56
2.	Copper Sulphate	25	---	---	---
3.	Cuprous Oxide	80	---	---	---
4.	Copper Hydroxide	---	---	50	---

Principle: The Copper oxy chloride which is insoluble in water is digested with 1:1 HCl (or conc. Nitric acid) to form water soluble cupric ions (i.e. $CuCl_2$) which when treated with Potassium iodide liberate one equivalent amount of Iodine. Thus liberated Iodine is estimated by standard Sodium thiosulphate solution using starch solution as indicator.

Reactions:

$$3Cu(OH)_2 \, CuCl_2 + 2HCl ---- 4CuCl_2$$
$$2CuCl_2 + 4KI ---- 2CuI + 4KCl + I_2$$
$$I_2 + Na_2S_2O_3 ---- 2NaI + Na_2S_4O_6$$

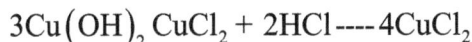

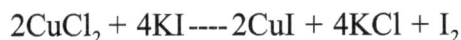

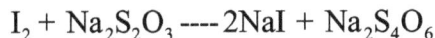

Calculations:

2 equivalents of Sodium thiosulphate = 2 equivalents of Iodine

= 2 equivalents of Copper

One equivalent of Sodium thiosulphate = 1 equivalent of Copper

= 63.5 gms of Copper

1000 ml of 1 N Sodium thiosulphate = 63.5 gms of Copper

't' ml of N normal Sodium thiosulphate = $\dfrac{63.5 \times t \times n \text{ gms.}}{1000}$

% Copper content = $\dfrac{63.5 \times t \times n}{W}$

Where,

- t = Volume in ml of Sodium thiosulphate consumed.
- n = Normality of Sodium thiosulphate.
- D = Dilution factor.
- W = Weight in gms of sample taken.

Procedure: Weigh accurately a quantity of sample which contains 0.2 gm of a.i in a 250 ml iodine flask, add 2 to 3 ml conc. Nitric acid, 20 ml distilled water, allow the material to dissolve by shaking, then boil for 3 to 5 minutes. Cool the flask, add 1 gm. of Urea and boil again for about 5 min. Then cool and add Sodium carbonate in small quantities with swirling until a faint precipitate or a blue colour appears. Add 10% acetic acid solution till the precipitate redissolves and solution becomes green. Add 5 ml of 30% KI solution and titrate the brown solution with 0.1N standard Sodium thiosulphate to a pale straw yellow colour. Add 1 ml of 1% starch solution, 1 to 2 gms of Potassium or Ammonium thiocyanate and continue the titration until the blue colour just discharged.

Precautions: The boiling or digestion of the sample with acid solution should be for sufficient duration to convert completely water insoluble Copper to water soluble Copper. Hence the boiling of the sample with acid may be done for longer duration if necessary.

Dithiocarbamates

Type of pesticide: They are protective and contact fungicides belonging to group of derivatives of dithiocarbamic acid.

Name	Molecular formula	Molecular weight	Structure
Alkyl dithiocarbamates			
1. Thiram	$C_6H_{12}N_2S_4$	240.4	
2. Ziram	$C_6H_{12}N_2S_4Zn$	305.8	
3. Ferbam	$C_9H_{18}FeN_3S_6$	418.9	
Alkylene dithiocarbamates			
1. Zineb	$C_4H_6N_2S_4Zn$	275.74	
2. Mancozeb	$C_4H_6N_2S_4MnZn$		
3. Propineb	$C_5H_8N_2S_4Zn$	290	
4. Metiram	$[C_{16}H_{33}N_{11}S_{16}Zn_3]_x$	$(1088.7)_x$	

Registered products:

S.No.	Name	% A.I in different products				
		Tech	WP	CS	DS	WS
1.	Ferbam	81	75	-	-	-
2.	Mancozeb	85	75	-	-	-
3.	Thiram	95 & 98	-	-	75	75
4.	Zineb	80	75	-	-	-
5.	Ziram	95	80	27	-	-
6.	Propineb	80	70	-	-	-
7.	Metiram	--	75	--	--	--

- Principle: One mole of dithiocarbamate on decomposition with hot boiling dilute acid liberates two moles of carbondisulphide (3 moles in case of Ferbam, 8 moles in case of Metiram) which, after removal of Hydrogen sulphide (produced from impurities) by absorption in Cadmium sulphate/Lead acetate solution, is converted quantitatively in methanolic KOH to Potassium xanthate. This Potassium xanthate is quantitatively estimated in acetic acid medium by titration against standard Iodine solution using starch solution as indicator.

- Reactions:

Step - l: Alkyl dithiocarbamates

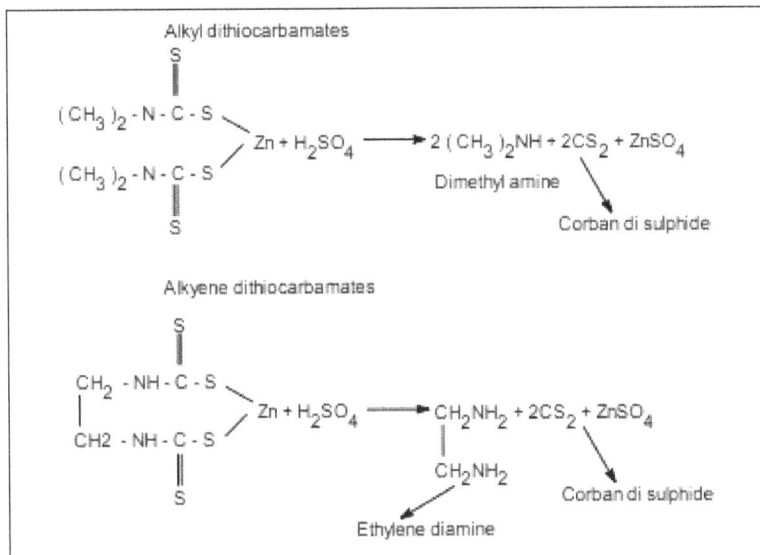

Step - II

$$CS_2 + CH_3OH + KOH \rightarrow CH_3O - C - S - K + H_2O$$

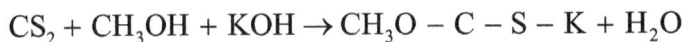

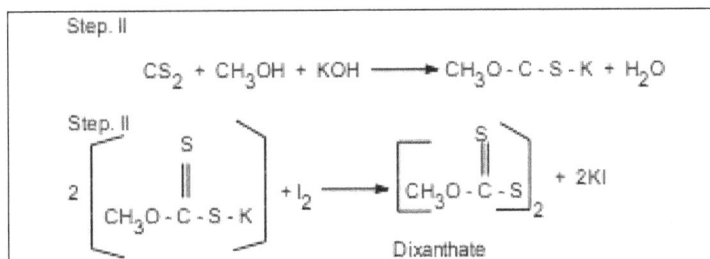

In the case of alkyl bisdithiocarbamates, Ziram, Thiram, Ferbam. A.I content can also be estimated on the basis of dimethylamine content.

Calculations:

a) Thiram, Ziram, Mancozeb, Propineb and Zineb:

2 equivalents of Iodine = 2 equivalents of CS_2 = 1 mole of a.i

1 equivalent of Iodine = ½ mole of a.i

1000ml of 1 N Iodine solution = ½ mole of a.i

't' ml of N normal Iodine solution = $\dfrac{\text{mol. Wt.} \times t \times n}{2 \times 1000}$ gms of a.i

Hence, % A.I = $\dfrac{\text{mol. Wt.} \times t \times n}{20W}$

Where,

- 't' = volume of standard Iodine solution consumed by the sample.

- N = normality of Iodine

- W = weight in gms of sample taken.

b) Ferbam:

3 moles of CS_2 = 1 mole of a.i

1 mole of CS_2 = 1/3 mole of a.i

1 equivalent of Iodine = 1 mole of CS_2 = 1/3 mole of a.i

Hence, % A.I = $\dfrac{\text{mol. Wt.} \times t \times n}{30W}$

c) Metiram:

8 moles of CS_2 = 1 mole of a.i

1 mole of CS_2 = 1/8 mole of a.i

1 equivalent of Iodine = 1 mole of CS_2 = 1/8 mole of a.i

Hence, % A.I = $\dfrac{\text{mol. Wt.} \times t \times n}{80W}$

- Procedure: Take 100 to 150 ml of 1 N Sulphuric acid in reaction kettle. Take 25 ml Lead acetate (Zinc acetate in case of Propineb) in first absorber and keep it hot (approximately 70 °C). Take 25ml 2 N methanolic KOH in the second absorber keep it chilled below 10 °C.

  Assemble the train of apparatus. Connect it to the vacuum line and maintain a regular flow of air through the assembly. Heat the acid in the kettle and keep it boiling. Transfer sufficient quantity of sample accurately by weighing 0.2 to 0.3 gms of a.i into the reaction flask. Digest for one hour and fortyfive minutes. After the digestion is complete, quickly disconnect the KOH trap. Transfer quantitatively with distilled water the contents of the trap in to a 500ml iodine flask. Neutralise it with 30% acetic acid using phenolphthalein indicator, add 1ml excess acid and titrate immediately with 0.1N Iodine solution using starch solution as indicator. End point is colourless to blue.

  In case of Thiram the acid used is acetic acid (1:1) and Zinc oxide mixture instead of 1.1N Sulphuric acid and 19.5% Cadmium sulphate solution is used instead of Lead acetate in the trap.

Precautions:

- The final titration of Xanthate against standard Iodine solution should be done very rapidly within two to three minutes because xanthate is unstable in neutral or faintly acidic medium.

- The sample should be brought in contact with only hot acid not with cold acid, because in cold acid only one and not two moles of carbon disulphide is released per mole of a.i.

Tridemorph

- Type of pesticide: Systemic fungicide belonging to morpholine group.
- Molecular formula: $C_{19}H_{39}NO$
- Molecular weight: 297.5
- Structure:

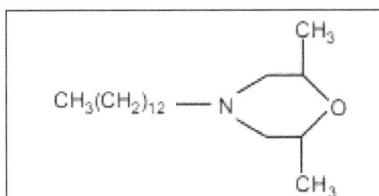

- Registered products:
  - Tech. 95%
  - EC 75%
- Method of Analysis: Acid base titration in non-aqueous medium.
- Principle: AI is titrated against perchloric acid in aqueous medium like acetic acid, when one gram equivalent perchloric acid is consumed b one gram mole of a.i.
- Reactions:

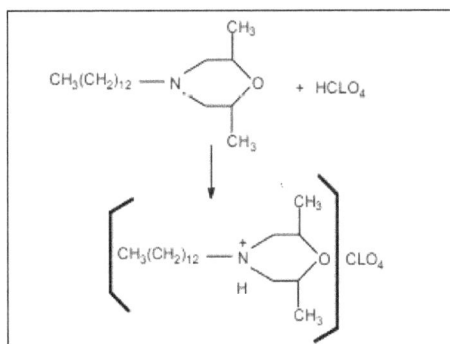

Calculations:

One-gram eqvt. of perchloric acid = 1 gm mole of a.i.

1000ml of 1N perchlorc acid = 297.5 gms. of a.i

Hence, % a.i $= \dfrac{29.75 \times t \times n}{W}$

Where,

- t = Titration reading in ml. of std. perchloric acid.
- N = Normality of perchloric acid.

- W = Weight in gms. of sample.

Procedure:

Weigh accurately the sample equivalent to 1 gm a.i in a flask. Dissolve in 50 ml acetic acid. Add 10ml acetic anhydride. Keep it for five minutes then titrate it against std. perchloric acid.

Phorate

- Type of pesticide: Systemic insecticide belonging to organophosphorus group.

- Molecular formula: $C_7H_{17}O_2PS_3$

- Molecular weight: 260.4

- Structure:

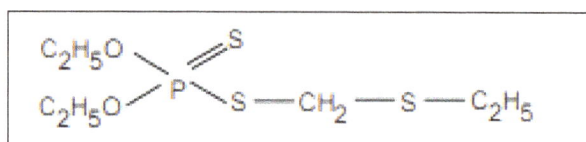

- Registered products: Technical 90% CG 10% Principle: One mole of a.i on hydrolysis in acidic medium yields one mole each of diethyl dithio phosphoric acid and ethyl mercaptan, which consume two equivalents of Silver nitrate. From the quantity of Silver nitrate consumed by a.i alone, obtained after correcting for impurity interferences, a.i content is calculated.

- Reactions:

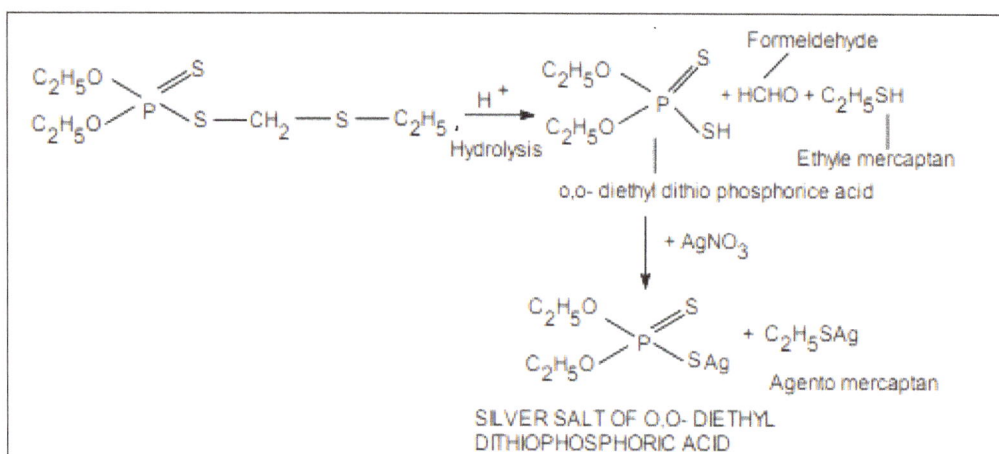

Calculations:

Gm equivalents of Silver nitrate = 1 gm mole of a.i

1gm equivalent of Silver nitrate = ½ mole of a.i = 130.2 gms

1000ml of 1 N Silver nitrate = 130.2 gms of a.i

't' ml of N normal Silver nitrate $= \dfrac{130.2 \times t \times n}{1000}$

Hence, % A.I $= \dfrac{13.02 \times t \times n}{W}$

Where,

- t = volume in ml of standard Silver nitrate solution consumed by a.i alone

- N = normality of Silver nitrate

- W = weight of sample taken

Procedure:

Step I: Preparation of sample stock solution

Sufficient quantity of the sample equivalent to 2.5 gms of a.i is extracted with acetone quantitatively and made up to 250ml in a volumetric flask.

Step II: Estimation of a.i and impurities

To a flask containing 150 ml water acidified with 5 drops of Nitric acid, add by pipette 50 ml of standard Silver nitrate solution and maintain at 50 °C. After attaining constant temperature, add by a pipette 50 ml of sample stock solution, and carry out hydrolysis at 50 °C for 15 minutes. Titrate the excess silver nitrate against standard thiocyanate solution using 1 ml 10% Ferric nitrate solution as indicator (x ml).

Step III: Estimation of impurities

Transfer by pipette 50ml of sample stock solution in to a 500 ml separating funnel containing 150ml enzene. Extract this solution thrice, each time by 50ml of 0.1N KOH in aqueous solution of 10% Potassium nitrate followed by one washing with 50ml distilled water. Collect the aqueous layer and acidify with concentrated Nitric acid till acidic to Congo red paper. Add 5 drops of acid in excess and keep it in waer bath maintained at500C till contents of the flask attained constant temperature. Then add by pipette 5 ml standard Silver nitrate and keep it in the water bath at 500C for 15 minutes. Then cool to room temperature and titrate the excess Silver nitrate against std. thiocyanate solution using 1 ml of 10% Ferric nitrate as indicator (y ml).

Calculate 't' value as shown below:

$$\text{'t'} = \left(50F_1 - xF_2\right) - \left(5F_1 - yF_2\right)$$

Where,

- F_1 = Ratio of normality of Silver nitrate to 0.1

- F_2 = Ratio of normality of thiocyanate to 0.1

Use a dilution factor of 5 while calculating final result.

Sulphur

- Type of pesticide: Contact and protective fungicide, acaricide

- Molecular formula: S

- Molecular weight: 32.06

- Registered products:

  ◦ Technical - 99.5%

  ◦ DP - 85%

  ◦ WP - 70% & 80%

  ◦ CS - 40%

  ◦ Solution - 22%

- Principle: Sulphur is converted to thiosulphate by refluxing with Sodium sulfite. Excess Sodium sulfite is destroyed by Formaldehyde and then the thiosulphate is estimated by titration with standard Iodine solution using starch solution (1%) as indicator.

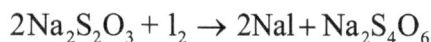

- Reactions:

$$S + Na_2SO_3 \rightarrow Na_2S_2O_3$$
$$2Na_2S_2O_3 + 1_2 \rightarrow 2NaI + Na_2S_4O_6$$

Calculations:

2 gm equivalent of Iodine = 2 gm mole of Hypo

= 2 gm mole of Sulphur

1 gm equivalent of Iodine = 1 gm mole of sulphur = 32.06 gms of Sulphur

1000ml of 1 N Iodine solution= 32.06 gms of Sulphur

't' ml of N normal Iodine solution $= \dfrac{32.06 \times t \times n}{1000}$

If dilution is made the, Sulphur content % by mass $= \dfrac{3.2\,06 \times t \times n \times d}{W}$

Outline of the procedure: Weigh accurately sufficient quantity of sample to contain 0.1 gm of Sulphur in a 250ml flat bottom flask add 30ml to 40ml distilled water, 2 gms of sodium sulfite and few drops of liquid paraffin. Reflux for 40 minutes, cool the flask and remove, add 10ml formaldehyde, 10ml acetic acid solution and titrate immediately with standard Iodine using freshly prepared Starch solution ('t' ml).

Carry out a reagent blank without sample as per above procedure.

Aluminium Phosphide

- Type of pesticide: Phosphine (fumigant) releasing product used in rodent control and stored grain protection

- Molecular formula: AlP

- Molecular weight: 57.96

- Registered products: Tablets 56%, 55%, 57%, 60% for export

- Reactions:

$$2AlP + 3\ H_2SO_4 \xrightarrow{\text{Nitrogen atmosphere}} Al_2(SO_4)_3 + 2PH_3$$

$$2PH_3 + 8(O) \longrightarrow 2H_3PO_4$$

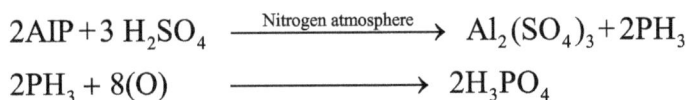

Calculations:

16 gm equivalents of Oxygen = 2 gm mole of Phosphine = 2 gms mole of a.i

1 gm equivalent of Oxygen = 1/8 gm mole o a.i = 7.24 gms of a.i

1000 ml of 1 N solution = 7.24 gms of a.i

't' ml of N normal solution $= \dfrac{7.24 \times t \times n}{W}$

Hence, % A.I $= \dfrac{0.724 \times t \times n}{W}$

Where,

- 't' = Quantity in ml of oxidant solution.

- n = Normality.

- W = Weight in gms of sample.

Procedure:

0.2 to 0.3 gms of material is decomposed with 10% Sulphuric acid in inert atmosphere. Phosphine liberated is swept by continuous stream of Nitrogen in to 200ml of standard 0.5N Potassium permanganate solution measured accurately and distributed in three different gas absorption bottles. Initially decomposition is carried out for 30 to 45 minutes at room temperature. Then the flask is heated on water bath maintaining at 60 °C for another 15 minutes. Transfer quantitatively permanganate solution into a 1 lit. beaker and add 200ml (sufficiently excess Oxalic acid) standard Oxalic acid (0.5N) solution. Titrate the excess Oxalic solution against standard Potassium permanganate solution.

$$\% \text{ A.I} = \frac{0.724 \times (200 + A)\ N_1 - (200\ N_2)}{W}$$

Where,

- A = Sample titration reading.
- N_1 = Normality of $KMnO_4$.
- N_2 = Normality of Oxalic acid.
- W = Weight of sample.

Precautions:

- Phosphine is highly toxic and inflammable gas. Hence decomposition should be carried out only in inert atmosphere.
- Introduce sample into reaction flask only after removing air by sweeping apparatus with Nitrogen gas for 15 minutes.

Zinc Phosphide

- Type of pesticide: Acute oral rodenticide.
- Molecular formula: Zn_3P_2
- Molecular weight: 258
- Registered products: Technical 80% min to 20% (for export) 1% & 2% (household products).
- Principle: One mole of a.i on acidic decomposition in an inert atmosphere yields two moles of phosphine which is oxidized by absorption in Potassium permanganate solution. From the quantity of permanganate solution consumed by phosphine, quantity of a.i is calculated.

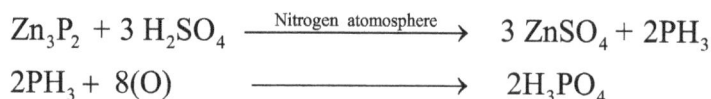

$$Zn_3P_2 + 3 H_2SO_4 \xrightarrow{\text{Nitrogen atomosphere}} 3 ZnSO_4 + 2PH_3$$
$$2PH_3 + 8(O) \longrightarrow 2H_3PO_4$$

Calculations:

16 gms equivalent of (O) = 2 gm mole of Phosphine = 1 gm mole of a.i

1 gm equivalent of (O) = 1/16 gm mole of a.i = 16.13 gms of a.i

1000 ml of 1 N solution = 16.13 gms of a.i

$$\text{t ml of N normal solution} = \frac{16.13 \times t \times n}{1000}$$

Hence, % A.I = $\dfrac{1.613 \times t \times n}{W}$

Where,

- t = Quantity in ml of oxidant solution.

- N = Normality of oxidant solution.

- W = Weight in gms of sample.

Procedure: Weigh accurately 0.5 gms of sample and analyse as per method outlined under aluminium phosphide described earlier. In order to remove adulteration by inorganic sulphides, phosphine carried by nitrogen is passed through 100ml of 1 N Sodium hydroxide solution before passing through Potassium permanganate solution.

Methyl Parathion

- Type of pesticide: Contact insecticide belonging to organophosphorus group.

- Molecular formula: $C_8H_{10}O_5NPS$.

- Molecular weight: 263.21.

- Structure:

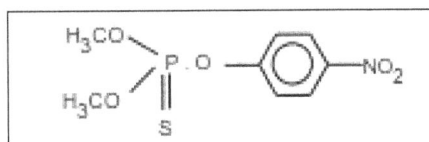

- Registered products:

  ○ Technical 92%

  ○ Technical concentrate 80%

  ○ DP 2%

  ○ EC 50%

- Principle: The phenolic impurities mainly para nitrophenol are separated from a.i by washing the ethereal sample solution with dilute chilled sodium carbonate A.I is reduced to amino derivative which is estimated by diazotization titration.

- Reactions:

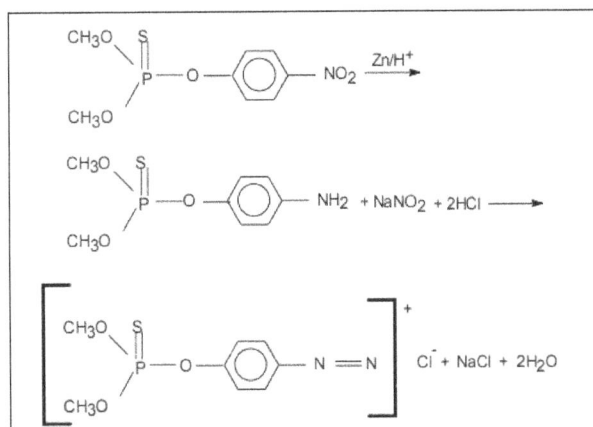

Calculations:

1 gm equivalent of Sodium nitrite = 1 gm mole of a.i = 263.21 gms of a.i

1000 ml of 1N Sodium nitrite = 263.21 gms of a.i

't' ml of n normal Sod. Nitrite = $\dfrac{263.21 \times t \times n}{1000}$ gms of a.i

Hence, % A.I = $\dfrac{26.321 \times t \times n}{W}$

Where,

- t = ml of standard solution of Sodium nitrite consumed by the sample.

- N = Normality of sodium nitrite.

- W = Weight of the sample.

Procedure: 1 gm a.i equivalent of sample is dissolved in 100 ml ether contained in a separating funnel and wash repeatedly, very carefully using each time 10 to 20 ml of chilled 1% Sodium carbonate solution till the aqueous solution is free from yellow colour. Then ether layer is taken in to a beaker and mixed with acid mixture (9:1 Acetic acid and HCl), add about 2 gms of Zinc dust or granules and heat it on water bath for about 1 hr. and the solution in the beaker is colourless. Destroy the excess Zinc by adding conc. HCl and heating. Cool and chill it to about 0oC, add 5 gms of Sodium bromide or Potassium bromide and titrate it slowly and carefully against standard Sodium nitrite solution using starch iodide paper as external indicator.

Physico-chemical Characteristics

Even though A.I content is the most important parameter in a formulation, there are certain other parameters which have profound influence on field bio efficacy of the formulation and hence are as much important as A.I content. These parameters are of physico-chemical nature. So testing of samples for A.I content alone is not sufficient measure of quality of the product. This does not reflect on field performance when used at any time within expiry date and hence is incomplete analysis. All samples have to be tested for physico-chemical parameters also after estimation of a.i content. Some of the important physico-chemical parameters are listed in table.

S. No.	Parameter	Application for
1.	Acidity/Alkalinity	All samples
2.	Particle size	DP/WP/Gr
3.	Bulk density	DP
4.	Suspensibility	WP
5.	Emulsion stability	EC
6.	Water runoff test	Coated Gr
7.	Attrition test	Coated Gr
8.	Flash point	EC

BIS has compiled and published methodologies for testing of these characteristics. For the purpose of training a beginner, simplified and condensed methodologies are given in this topic.

Acidity and Alkalinity

- Preliminary qualitative test: Mix 0.5gms or ml of sample with 1ml water and find out whether the sample is acidic or alkaline by using litmus or pH paper.

- Quantitative analysis:

  ○ For liquid formulations: 10gms sample, dispersed or dissolved in 100 ml of water is titrated against 0.05 N Sodium hydroxide or HCl using either bromo cresol purple (1% alcoholic solution) methyl red (1% aqueous solution).

    Alternatively one may determine the end point using pH meter in which case end point is pH of mixture of 50 ml of acetone and 5 ml of standard buffer solution (100 ml of 2N acetic acid 100 ml of 1N Sodium hydroxide).

- For solid formulation: 10 gms of sample is mixed with 25 ml acetone and warmed gently on a hot water bath. Cool, dilute it with 75 ml water and allow it to stand for 1 hour with intermittent agitation. Decant or filter the liquid and titrate.

Calculations:

$$\text{Acidity (as \% Sulphuric acid)} = \frac{4.9 \times t \times n}{W}$$

$$\text{Alkalinity (as Sodium hydroxide)} = \frac{4.0 \times t \times n}{W}$$

Particle Size

- DP: 10gms of sample is weighed and transferred to a test sieve (75 micron or 200 mesh) and shaken in a rotap shaker for 20 min. Material passing through or retained on the sieve is weighed.

- WP: 10gms of sample is mixed with 100ml of water, allowed to stand for half minute and agitated gently with glass rod for half minute. Transfer the slurry on to sieve (75 or 45 micron as required) rinsing with tap water. Pass running water till all solids pass through. Weigh the material retained on the sieve after drying in a tarred gooch crucible.

- Granules: 100gms of sample is sieved for 15 min. through set of sieves of upper and lower particle size declared on the label.

- Calculations:

$$\text{Material passing through the sieve} = \frac{W1 - W2 \times 100}{W1}$$

Where,

- W1 = Weight in gm of sample.

- W2 = Weight in gm of sample retained on the sieve.

Bulk Density

- Find the weight of 100ml of sample filled in a measuring cylinder and calculate the density D^1.

- Stopper and gently tap or drop the cylinder 20 times on a felt pad or any other soft surface through a height of 15 cm. Measure the volume and calculate the density D^2.

- Requirement: D^2 should not be more than $(D^1 + 0.6D^1)$.

Suspensibility

Percentage of A.I held in the suspension of known concentration prepared in standard hard water, after keeping for 30 minutes to stand for sedimentation if any.

Step – I: Preparation of suspension

Weigh in a beaker enough quantity of sample to give (on mixing with water) 250ml of suspension containing 0.5% or 2.5% a.i as per relevant specification requirement. Add about 50ml of standard hard water of 342 ppm. Of Calcium carbonate (Dissolve 304 mgs of Calcium chloride and 139mgs of magnesium chloride in 1 lit. of water) and stir gently with glass rod. Transfer the slurry quantitatively in to a 250ml Stoppard measuring cylinder using standard hard water and diluting it up to the mark, Stopper the cylinder and turn it quickly (30 times in a minute end over end. Allow the suspension to settle for 30 minutes.

Step – II: Sucking the suspension

Gently remove by mild suction 225ml of suspension without disturbing bottom 25ml (within 15 seconds) and discard it.

Step – III: Estimation

Transfer the bottom 25ml of suspension/sediment and estimate A.I content by relevant method given in the specification.

Step – IV: Calculation

$$\% \text{ Suspensibility} = \frac{1000 \left(W^1 - W^2 \right)}{9 \, W^1}$$

Where,

- W^1 = Weight in gm of a.i taken for test.

- W^2 = Weight of a.i estimated in the bottom 25 ml sediment.

Emulsion Stability

2 or 5 ml of sample is emulsified with standard hard water and diluted to 100 ml in a Stoppard measuring cylinder. Allow it to stand for one hour on a vibration free bench. Observe the emulsion for creaming at the top and sediment at the bottom and measure the same, which should not exceed 2 ml.

Water Runoff Test

Soak 10 gms of sample for 15 min in 50 ml of water contained in a burette or glass tube fitted with a stopcock. After the specified time drain off the water and estimate the a.i contained in the water.

Attrition Test

Determine the a.i in the residue on 75 micron sieve after shaking 100gms of sample for 2 hours on rotap sieve shaker.

Flash Point

Fill the cup (sample holder) of Abel's closed cup flash point apparatus with the sample up to the recommended level. Close the cup. Set up the test flame as very small dot or button. Measure the temp. of the sample. Now introduce the test flame in to empty space above the surface of sample in the closed cup and check whether the vapour contained in the empty space flashes or not. By repeating the above process at different temperatures (by either heating or cooling the sample). The flash point of the sample should be above 24.5 °C.

Spectroscopy

Spectroscopy is the science related to the study of interaction of radio magnetic energy with the atoms, ions or molecules. When the electromagnetic radiation falls on the matter part of the radiant energy is absorbed and remaining is transmitted are scattered or reflected. If the scattered or reflected radiation is kept at least and constant, the measurement of transmitted radiation to that of incident radiation is possible and the amount of absorbed radiation can be deduced out of it. The relationship can be plotted to obtain spectrum called absorption/transmission spectrum. This type of study is known as molecular absorption Spectroscopy. The instruments used for this study are Spectrophotometers.

The excited atoms, ions or molecules on irradiation with a beam of electromagnetic radiation, relax to lower energy levels by giving up their excess energy as photons (fluorescence or phosphorescence) to attain stability and such emitted radiant energy can be measured and plotted as a spectrum which is called emission spectrum and the study the emission spectroscopy. The instruments used for this study are Fluorometers and Spectrofluorometers.

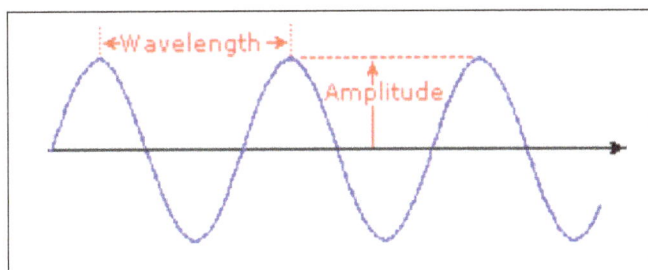

Important terminology used:

- Wave length: maybe defined as distance between two troughs or crests of a wave (denoted as lambda) and expressed in micron, millimicron/nanometer (nm).

- Micron: 10^{-4} cm.

- Millimicron/Nanometer: 10^{-7} cm.

- Wave number is defined as number of waves per centimeter. 1/wavelength.

- Frequency may be defined as number of waves that passes through a unit in unit time.

- $V = C/\lambda$.

- Velocity of light: $C = 3 \times 10^{10}$ Cm/sec.

- E = Energy.

- Planks constant = h = 6.623×10^{-27} erg sec/mol

- Monochromatic light: The light which consists of polychromatic/multi wavelength radiation is resolved using a prism or diffraction grating to give single wavelength radiation which is called monochromatic light.

The electromagnetic radiation consists of a broad spectrum of radiations which are named on the basis of their wave length or frequency as Gamma, X rays, Ultraviolet, Visible, Infrared rays and Hertzian, Radio waves and Audible frequencies.

The radiation which are important for absorption spectroscopy are tabulated here under.

S. No	Type of radiation	Wave length	Energy
1.	Ultra violet (UV)	200 to 375 nm	More
2.	Visible (Vis.)	400 to 760 nm	Medium
3.	Infrared (IR)	1 micron to 50 micron	Less

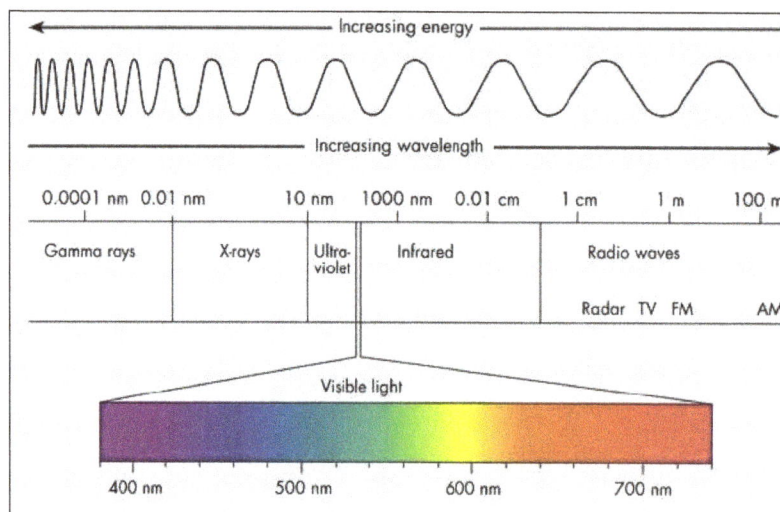

Every radiation has got a defined amount of energy which is given by the relationship E = hC/wavelength, where C is the speed of light and h planks constant whose product is constant and hence the energy is associated with each wave length is unique and characteristic of it. The type of change that is brought about in the energy level of the molecule by each monochromatic light is specific as that of their energy. They may be listed as follows:

S. No	Type of radiation	Type of energy level change
1.	Ultra violet (UV)	Change in vibrational and rotational energy level.
2.	Visible (Vis.)	Change in electronic level leading to excitation.
3.	Infrared (IR)	Change in electronic level leading to excitation.

- Application: The molecular absorption spectroscopy can be applied for the purpose of qualitative and quantitative analysis.

- Qualitative Analysis: As each molecular orbital in the molecule require a specific energy to excite to the higher energy level, it absorbs a specific electromagnetic radiation or monochromatic light and becomes the qualitative parameter for identification. Usually it is the lambda maximum and the shape of the spectrum of the unknown compared with a known pure reference standard spectrum or compared with the spectra available in the literature for its identification.

- Quantitative analysis: The quantitative analysis is based on two laws namely (i) Lamberts law and (ii) Beer's law and the absorption is proportional to the concentration of the solute present in the solution within the limits of the above laws.

- Lamberts law: States that when monochromatic light passes through a transparent medium, the in intensity of the emitted light decreases exponentially as the thickness of the absorbing medium increases arithmetically, or that any layer of given thickness of the medium absorbs the same fraction of the light incident upon it.

- Beer's law: State that the intensity of a beam of monochromatic light decreases exponentially as concentration of absorbing substance increases arithmetically.

Combining both the laws we have Beer-Lamberts law,

$$A = ecl = \log I_0/I_t = \log 1/T = -\log T$$

Where,

- A = Absorbance.
- e = Molar absorption coefficient (when concentration is expressed in mole dm^{-3}).
- c = Concentration.
- l = Path length.

For any given molecule the molar absorption coefficient is constant and if a constant path length cell is used then A is proportional to the concentration. And this principle is used for quantitative estimation of the substances of known by comparing with the response/absorbance of a pure reference standard in a single point calibration method or can draw a calibration curve of the reference standard and read out the unknown concentration from its absorbance.

Deviations from the Beer's law:

- It is most reliable when radiation is monochromatic.

- It is valid or linear relationship is observed only up to a particular concentration. Hence quantitative analysis should be done within this range of linearity validity.

- Solutions should be absolutely clear, even trace quantity of fine particles floating in solution will give erratic results.

- The cells should have absolutely transparent surfaces because Beer-Lamberts law assumes minimum reflection of light.

UV-Visible Spectrophotometer

The UV-Visible Spectrophotometer consists of mainly a source, monochromator, Beam splitter, sample path and reference path, cell holders and cells and the detector. The signal from the detector is fed to an internal microprocessor or external personal computer, which will analyse the data, present in the desired spectral form and generate the reports.

The schematic diagram of the UV-Visible Spectrophotometer is as follows:

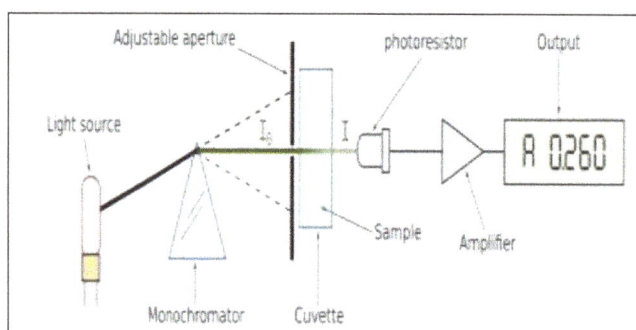

Single beam spectrophotometer.

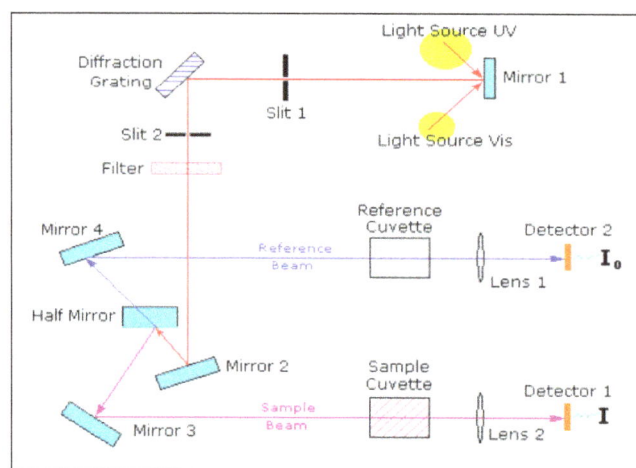

Double beam spectrophotometer.

- Source of light: The UV-Visible Spectrophotometer is equipped with two types of sources to give the radiant energy to cover both the ranges of UV and Visible as no single source will be able to cover both the ranges. The deuterium lamp is provided to supply the UV range of light, whereas the tungsten lamp provides the visible range. Both the lamps are aligned in such a way to give continuous spectrum of UV and Visible range and the change over takes place automatically without any hindrance in scanning and recording the complete range from 200 nm to 1100 nm.

- Monochromator: Earlier days the colorimeters were manufactured using filters as monochromators and need to be changed depending on the requirement of the sample under investigation and the data arrived are not so accurate.

Wavelength of absorbance maximum (nm)	Color Absorbed	Color Remaining
380-420	Violet	Green-yellow
420-440	Violet-blue	Yellow
440-470	Blue	Orange
470-500	Blue-green	Red
500-520	Green	Purple
520-550	Yellow-green	Violet
550-580	Yellow	Violet-blue
580-620	Orange	Blue
620-680	Red	Blue-green
680-780	Purple	Green

The modern spectrophotometers are manufactured using either a prism or dispersive grating as the monochromator, which will resolve spectrum to give an accuracy of 0.1nm. The monochromator is made to move on rotating base with the help of a motor to project the required wave length through the slit.

- Beam splitter: The spectrophotometers are manufactured as single beam and double beam instruments. The single beam instruments are used in routine testing facilities where much accuracy is not involved and they are less expensive than the double beam spectrophotometers. The double beam spectrophotometers employ a design where the monochromatic light beam is split in to two and allowed to pass through two paths one for reference solution and another for the sample solution. The two beams are made to reach the detector simultaneously by employing a chopper, resulting in accurate measurement of the transmission or absorption of light that takes place in the sample solution.

- Detector: The most popular detector used in the Spectrophotometers is photomultiplier tube, which converts the energy of the photons which falls on it in to an electrical energy and fed to the amplifier. The amplifier transmits the data though the microprocessor connected to it which displays the transmission or absorption values or it can plot the spectrum in case of scanning.

- Photo Diode Array detector: The latest of the detector is the Photo Diode Array detector which simultaneously collects the data points throughout the scanning wave lengths and process electronically to give the spectrum within no time. This enables to identify the unknowns easily with much faster rate and can simultaneously obtain the spectrum and absorption data for different wavelengths in a single run. This particular detector is employed successfully in HPLC to monitor the unknowns where the quantity of sample available is much less and in dynamic state.

Dichlorophenoxy Acetic Acid (2,4-D)

- Type of pesticide: Selective translocated herbicide belonging to phenoxy group

- Molecular formula: $C_8 H_6 O_3 Cl_2$

- Molecular weight: 221

- Structure:

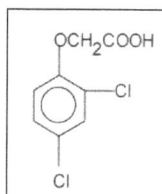

- Registered products:

S. No.	Product	Acid Equivalent	Active Ingredient
1.	2,4-D Technical	97%	97% (As acid)
2.	Ethyl Ester of 2,4-D – Technical	90%	80% (As acid)
			6% Free Acid (Max)
3.	Granules	4%	4.5% (As ester)
4.	W.P	18.5%	20.9% (As ester)
5.	E.C	3%	38.3% (As ester)
6.	SL	58%	69.8% (As salt)
7.	SP	80%	88% (As salt)

- Principle of analysis: Ester is converted to sodium salt. Salts (Sodium and amine salt) are acidified to release 2,4-D acid which is extracted, washed and is titrated against standard sodium hydroxide solution.

- Reactions:

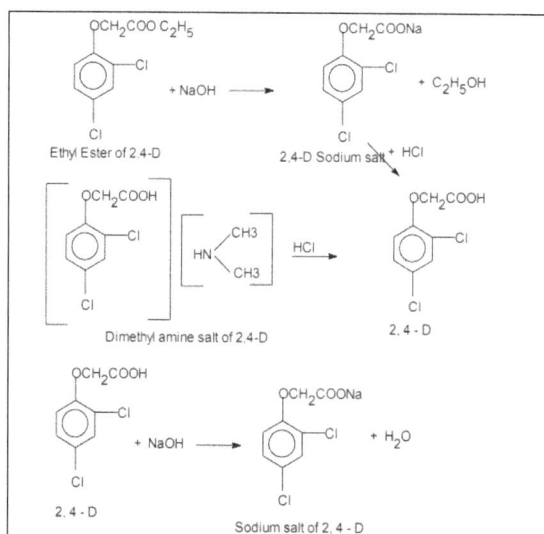

Calculations:

1 GE of NaOH = 1 gm of 2, 4-D

1000 ml of 1 (N) Na OH = 221 gm of 2, 4-D

$$\% \text{ AI as} = \frac{2.21 \times t \times n}{W}$$

Where,

- t = Titration reading in ml.

- N = Normality of NaOH.

- W = Weight of sample in gms.

$$\text{Ester content } \% = 2,4\text{-D} \times \frac{249}{221}$$

$$\text{Dimethyl amine salt content } \% = 2,4\text{-D} \times \frac{65.9}{221}$$

$$\text{Sodium salt content } \% = 2,4\text{-D} \times \frac{243}{221}$$

Introduce suitable dilution factor, if necessary, depending upon procedure.

Outline of the procedure:

- Technical 2,4-D

    ◦ A.I content: 5 gms of the sample, weighed accurately is dissolved in 50 ml of neutralized alcohol. Add 50 ml water and titrate against standard 1(N) NaOH using bromothymol blue (0.04% in alcohol) as indicator.

    ◦ Determination of free 2,4-D phenols:

        ▪ Preparation of standard solution: Dissolve 100 mg of pure 2,4-Dichlorophenol in 10 ml. acetone and dilute to 500 ml with water. Pipette 10 ml and dilute to 100 ml with water.

        ▪ Preparation of sample solution: Weigh 200 mg of sample and dissolve in 60 ml of standard 0.1(N) NH_4OH and dilute to 500 ml with distilled water.

        ▪ Colour development: Pipette 50 ml of sample solution; add 0.5 ml of standard 4-amino-phenazone solution (2% in water) and 0.25 ml of potassium ferricyanide (7% in water) shake well.

Pipette 10 ml of standard solution, add 5 ml of standard ammonium hydroxide solution, (0.1N) dilute to 50 ml with water. Add 0.5 ml of standard 4-amino-phenazone and 0.25 ml of potassium ferricyanide.

Measure absorbance in spectrophotometer.

- Amine salt of 2,4-D: 1 gm of 2,4-D equivalent sample is weighed accurately, dissolved in 10 ml wate, acidified with dilute Hcl (1:1), precipitate of 2,4-D extracted thrice with ether (25 ml each time), ether layer washed with minimum quantity of water till free from chloride ions, ether distilled off, residue dissolved in 50 ml neutral alcohol and 50 ml water and titrated against standard 0.1(N) NaOH using bromothymol blue as indicator.

  ○ Sodium Salt: A stock solution of the sample is prepared by dissolving 10 gm in 250 ml water. Pipette out 25 ml into separating funnel, acidify with 1.5 ml of dilute HCl (1:1), extract the acid with ether, distill off the ether and titrate the residue of 2,4-D dissolved in 60 ml netutral alcohol against 0.1(N) NaOH using broothymol blue as indicator.

  ○ Ester: About 1 gm of 2,4-D equivalent sample is refluxed for 1 hour with 1.5 gm KOH, 80 ml isopropanol and 20 ml water.

If the sample is EC: then extract the refluxed solution twice with petroleum ether. Then aqueous solution is first neutralized with few drops of HCl (1:1) using phenolphthale-in and then made faintly alkaline with NH_4OH (1:1). Then add 3 ml of 0.1% $BaCl_2$. Shake continuously; dilute it to 250 ml mark in a volumetric flask. Filter and pipette 100 ml into a separating funnel for further analysis.

If the sample is technical: then transfer the entire solution after reflux into a separating funnel. Acidify the solution with 3 ml HCl (1:1) extract 2,4-D with ether and proceed.

Identity Test

- For 2,4-D (Acid By TLC): Spot on a silica gel-G (250 micron) plate, 10 microlitres of 5% solution, develop with chloroform acetic acid (95:5). Dry the plate at 1000c for 20 minute. Spray with bromocresol green solution (0.04% in alcohol).

- For 2,4-D (Ester): GLC method.

- For amine salt: 1 ml of sample is mixed with 5 ml of 1% sodium nitro prusside 2.5 ml of 10% Acealdehyde and 5 ml of 2% sodium carbonate shake well. A dark blackish to brownish colour is produced after 5 minutes.

Carbaryl

- Description: Contact insecticide belonging to carbamate group.

- Molecular Formula: $C_{12}H_{11}O_2N$

- Molecular Weight: 201

- Structural Formula:

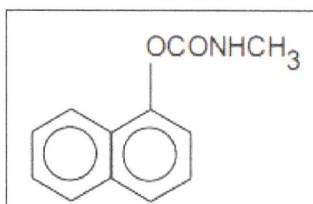

- Registered Products:
  - Technical - 97%
  - DP - 5% / 10%
  - WP - 50% / 85%
  - Flowable - 42%
  - Granule - 4%
  - LV - 40%
  - Combination - 4% each
  - Production (Carbaryl + Lindane)
- Method of Analysis:
  - Visible spectroscopy
  - HPLC for estimation of Beta carbonyl.

Visible Spectroscopy

- Principle: Carbaryl, on alkaline hydrolysis, yields one mole each of 1-Naphthol, methylamine, and carbondioxide per mole of A.I. The reaction product, 1-Naphthol, is reacted with iodine solution to yield purple red colour which is measured at 540 nm. The colour intensity (Absorbance) of test sample is compared with that of standard reference sample and % AI content in test sample is calculated.

- Reactions:

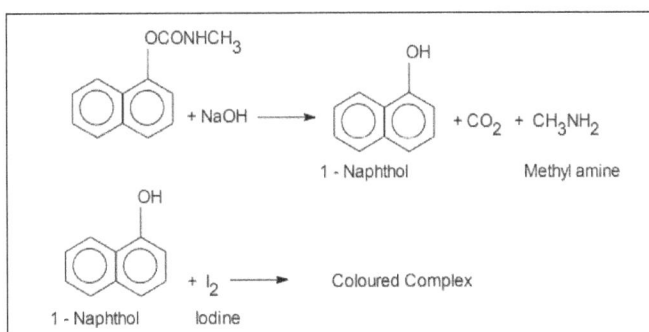

- Method:
  - Weigh, in two different flasks, enough quantity of reference standard and test sample equivalent to 60 mgm A.I.
  - Add 10 to 15ml methanol followed by 1 ml of 0.5 (N) potassium hydroxide solution in methanol and boil it for about 2 to 3 minutes. Add few more ml of methanol, if required, to prevent drying up of flask.

- ○ Cool it to room temperature and dilute the solutions to 100 ml in a volumetric flask, filtering if necessary, with methanol.

- ○ Pipette out 5 ml of these two solutions into two different 500 ml volumetric flasks, add about 300 ml water, followed by 20 ml of 0.01 (N) aqueous iodine solution. Shake well and dilute upto the mark. Prepare a reagent blank.

- ○ Measure absorbances of these two solutions at 540 nm against reagent blank.

- ○ Calculations:

$$\text{AI content \% (m/m)} = \frac{W1}{A1} \times \frac{A2}{W2} \times P$$

Where,

- W1 = Weight in gms of reference standard.
- A1 = Absorbance of reference standard.
- A2 = Absorbance of test sample.
- W2 = Weight in gms of test sample.
- P = Purity of reference standard.

Monocrotophos

- Description: Systemic insecticide belonging to Organophosphorus group.
- Molecular Formula: $C_7H_{14}O_5NP$.
- Molecular weight: 223.
- Structural Formula:

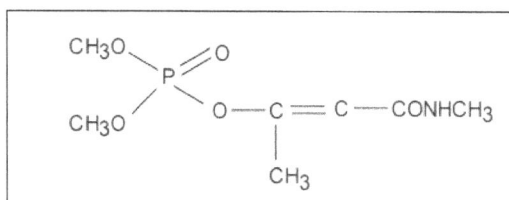

- Registered Products: Technical concentrate – 68% / 74%, SL – 36%
- Method of Analysis: Visible spectroscopy.
- Principle: Monocrotophos, on alkaline hydrolysis, yields one molecule each of 0,0-Dimethyl phosphoric acid, aceto acetic acid monomethyl amide (MMA) forms. After acidifying, MMA is reacted with Ferric ions to yield purple red colour which is measured at 544 nm. The colour intensity (Absorbance) of test sample is compared with that of standard reference sample of Monomethyl Amide of Aceto Acetic Acid (MMA), and % AI content in test sample is calculated. Thus, obtained value is corrected for interfering impurity in the test sample, namely free MMA, in test sample, estimated separately without hydrolysis of the sample.

- Reactions:

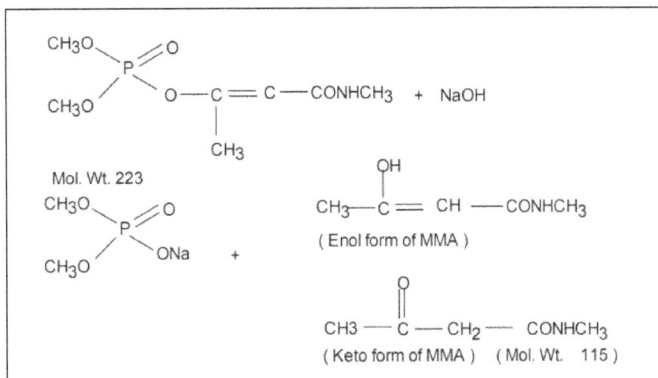

Method

Estimation of Total MCP Content

- Processing of sample:

  ○ Weigh enough quantity of sample equivalent to 2.9 to 3.0 gms of AI into a 100 ml volumetric flask, dissolve it in methanol and dilute it upto mark with methanol.

  ○ Transfer 10 ml of this solution to 250 ml volumetric flask and add 10 ml of 5 (N) aqueous sodium hydroxide solutions. Keep it for 30 minutes at room temperature.

  ○ Neutralise the free alkalinity with 1(N) HNO_3 and dilute upto the mark.

- Preparation of MMA standard solution: Weigh 0.15 gms of MMA and dilute it with methanol in a 250 ml volumetric flask.

- Colour development: Transfer 10 ml of solution A and B above into two 100 ml volumetric flasks. Add to both flasks 10 ml of 5% methanolic acetic acid, 50 ml methanol and 10 ml of 4% aqueous ferric chloride solution. Then add only to the flask containing MMA standard solution; 10 ml of sodium nitrate solution (prepared by neutralizing 10 ml of 5(N) NaOH with 1(N) HNO_3 and diluting to 250 ml in a volumetric flask).

  Dilute the contents of both flasks with methanol upto the mark and measure absorbance at 544 nm against reagent blank.

Calculation of total MCP content:

$$\% \text{ MCP (TOTAL)} = \frac{W1}{A1} \times \frac{A2}{W2} \times P \times 19.41 = (X)$$

Estimation of Free MMA

- Transfer adequate volume (usually 2 to 4 ml) from stock solution of test sample into a 100 ml volumetric flask. Develop colour as before and measure absorbance at 544 nm against reagent blank.

- Calculate MCP equivalent to free MMA impurity in the sample as shown below: MCP equivalent to free MMA.

$$\frac{W1}{A1} \times \frac{A3}{W2} \times P \times 0.776 = (Y)$$

Final Calculation:

% A.I (w/w) in test sample = (X) - (Y)

% Free MMA (w/w) in test sample = $\dfrac{Y}{1.94}$

Additional points:

- Reference standard to be used is MMA and not technical monocrotophos because with the latter, it is not possible to estimate free MMA.

- Technical MMA is highly hygroscopic. Hence it should be weighed quickly to minimize absorption of moisture. Alternatively, aqueous solution of MMA of known concentration may be used as reference standard.

- It is the Enol form of MMA which produces colour with ferric ion. Keto form of MMA does not produce colour with ferric ions. Hence about 10 minutes time should be given after addition of ferric ion before measuring absorbance. This will enable complete conversion of Keto to Enol form failing which erratic results will be obtained.

- While determining free MMA in the sample, absorbance values of standard and sample should be as close as possible. If not, volume of stock solution of the sample, to be taken may be proportionately reduced.

Malathion

- Description: Contact insecticide belonging to Organophosphorus group.
- Molecular Formulation: $C_{10}H_{19}O_6PS_2$
- Molecular Weight: 330
- Structure Formula:

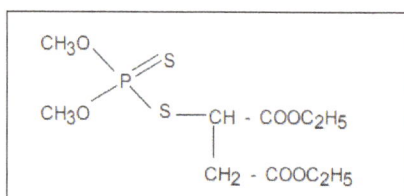

- Registered Products:
  - Technical - 95%
  - DP - 5%

  ◦   WP - 25%

  ◦   EC - 50%

  ◦   ULV - 96%

- Method of Analysis: Visible Spectroscopy.

- Principle: Malathion, on alkaline hydrolysis, yields one mole each of 0, 0-Dimethyl Dithiophosphoric acid, fumaric acid and two moles of ethanol per mole of AI. After acidifying, 0,0-Dimethyl Dithiophosphoric acid (DMTA) is complexed with cupric ions to yield a yellow coloured complex which is extracted with carbon tetrachloride and colour intensity (Absorbance) is measured at 420 nm. The absorbance value of test sample is compared with that of standard reference sample and % AI content in test sample is calculated.

- Reactions:

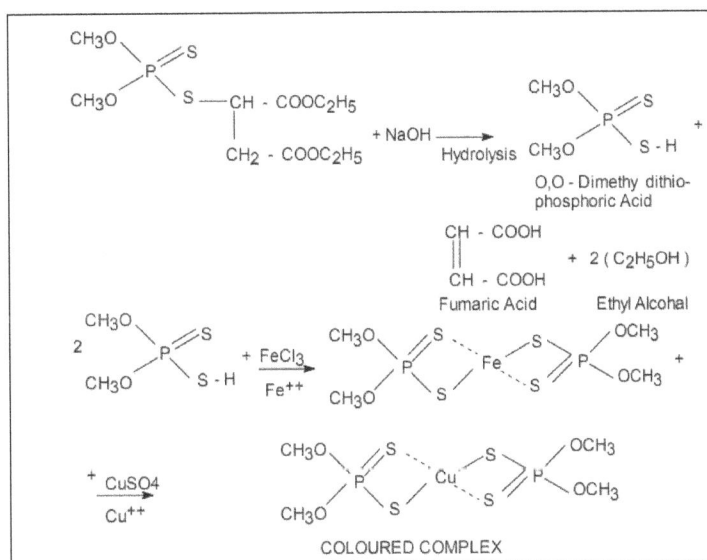

COLOURED COMPLEX

- Precautions: Coloured complex is unstable and hence absorbance of sample and standard should be measured at the same time lapse after addition of cupric ions.

Method:

- Weigh enough quantity of reference standard and test sample to contain 100 mgm AI. Dilute to 100 ml with ethanol/methanol in 100 ml volumetric flasks.

- Transfer 10 ml of these two solutions into two 100 ml volumetric flasks, add 1 ml acetonitrile and dilute upto the mark with methanol.

- Transfer 15 ml of these two solutions into two different 250 ml separating funnels. Add 2 ml of aqueous 0.5 (N) sodium hydroxide solutions. Allow the reaction (Hydrolysis) to proceed at room temperature for 2 minutes.

- Add 75ml of ferric reagent and keep it for 5 minutes.

- Add, either by pipette or by volumetric flask, exactly 50 ml of carbon tetrachloride.

- Add, by fast delivery pipette, 2 ml of 1.5% aqueous copper sulfate solution. Shake vigorously for 1 minute. Measure within 5 minutes absorbance of yellow coloured carbon tetrachloride layer at 420 nm against solvent.

- Calculations:

$$\text{AI content \% (W/W)} = \frac{W1}{A1} \times \frac{A2}{W2} \times P$$

Where,

- W1 = Weight in gms of reference standard.

- A1 = Absorbance of reference standard.

- A2 = Absorbance of test sample.

- W2 = Weight in gms of test sample.

- P = Purity of reference standard.

DIFFERENT TERMINOLOGIES USED IN PESTICIDE RESIDUE ANALYSIS

Pesticide Residue

Any substance or mixture of substances in food resulting from the use of a pesticide and includes any specified derivatives such as degradation and conversion products, metabolites, reaction products, and impurities that are considered to be of toxicological significance. It is express in mg/Kg.

Metabolism

Sum total of all physical and chemical process that take place within an organism; chemical changes that occur for a pesticide within an organism. It includes uptake and distribution within the body, changes (biodegradation) and elimination of pesticides and their metabolites.

Metabolite

After the pesticide enters the plant or animal body, it is normally converted into parts.

Dissipation and Persistence

The first step is the initial phase in which the disappearance of the residue is fast. This phase is called "Dissipation". The second phase, in which there is a slow decrease in the amount of residue, is known as "persistence".

No Observable Adverse Effect Level (NOAEL)

It is the highest dose of substance that does not cause any detectable toxic effects in experimental animal studies. It is expressed in mg/kg of body weight per day. To determine NOAEL values the sub chronic and chronic studies are carried out in different species as given under.

Sub Chronic

90-Day feeding study:

- Rodent (Rat, Mouse),
- Non-rodent (Dog),
- Dermal (Depending on Use Pattern),
- Inhalation,
- Neurotoxicity,

Chronic

One or Two-year oral study:

- Rodent (Usually Rat),
- Non-rodent (Dog),
- Life-time Oncogenicity Study,
- Reproductive,
- Multi-generation (Rat, Mouse), Fertility, Reproduction,
- Teratogenicity (Rat, Mouse, Rabbit),
- In-Vitro Mutagenicity and Mechanistic Studies.

Acceptable Daily Intake (ADI)

It is an estimate of the amount of a pesticide, expressed on a body weight basis that can be ingested daily over a lifetime without appreciable health risk.

$$\text{ADI for human beings (mg/kg body weight)} = \frac{\text{NOAEL animal studies (mg/kg body weight)}}{100 \text{ (safety factor)}}$$

Pre Harvest Interval

The pre harvest interval is the time interval from the application of pesticide to harvest under good agricultural practice so as to see that the pesticide residue falls below the MRL value.

Maximum Residue Level (MRL)

Is the maximum concentration of a pesticide residue in or on a food, agricultural commodity or animal feed, resulting from the use of a pesticide according to Good Agricultural Practice (GAP). The concentration is expressed in milligrams of pesticide residue per kilogram of the commodity. Under the PFA Act, MRL or Tolerance Limits (TLs) are fixed considering MRLs based on supervised trials conducted in India as well as the dietary habits of our population.

Toxicity studies of pesticides in animal are carried out to determine No observed Adverse Effect Level (NOAEL). Data are evaluated and NOAEL is calculated from chronic study. It is usually expressed in terms of milligrams of that particular pesticide per kilogram of body weight. From this NOAEL, the Acceptable Daily Intake (ADI) is calculated by dividing with a safety factor of 100. Therefore ADI, which is expressed in terms of mg/kg body weight, is an indication that if a human being consumes that amount of pesticide every day, throughout his lifetime, it will not cause appreciable health risk on the basis of well-known facts at the time of the evaluation of that particular pesticide. ADIs are derived from the results of long term feeding studies with laboratory.

MRL is therefore a dynamic concept dependant on extant knowledge and is therefore required to be renewed from time to time.

Terminal residues of a particular pesticide on a treated crop are estimated from supervised trials, to assess the maximum residue limit which the pesticide leaves when used as per the Good Agricultural Practice (GAP).

Thus, the above three parameters i.e. ADI, terminal residues as per Good Agricultural Practice on the crop and the diet pattern of the population are the critical inputs needed to derive the maximum residue limits (MRLs) of pesticides in food commodities.

BENEFITS OF PESTICIDES

- Pesticides help farmers to produce more with less land: With the introduction of pesticides, farmers have been able to produce bigger crops on less land, increasing crop productivity by between 20 and 50 percent. In addition, pesticides allow farmers to maximize the benefits of other valuable agricultural tools, such as high quality seeds, fertilizers and water resources. Pesticides are therefore an indispensable tool for the sustainable production of high quality food and fibers.

- Pesticides ensure bountiful harvests: Numerous scientific studies show that eating fruit and vegetables regularly reduces the risk of many cancers, high blood pressure, heart disease, diabetes, stroke, and other chronic diseases.

- Pesticides help keep food affordable: Farmers grow more food on the same land with the help of pesticides. Studies have shown that growers of organic vegetables spend significantly more on hand weeding compared to growers who use herbicides. This explains why organic food is more expensive than conventionally grown food.

- Pesticides help reduce waterborne and insect transmitted diseases: Such as malaria, Lyme disease and West Nile virus. Pesticides contribute to enhanced human health by preventing disease outbreaks through the control of rodent and insect populations.

- Pesticides help conserve the environment: They enable farmers to produce more crops per unit area with less tillage, thus reducing deforestation, conserving natural resources and curbing soil erosion. Pesticides are also critical for the control of invasive species and noxious weeds.

- Herbicides have removed the hardship of hand weeding: This means farming families across the world have the choice to pursue education and opportunities away from farming, thus improving quality of life and living standards.

- Pesticides have transformed developing countries into food producers: Crop protection products have helped farmers in the developing world grow two or three crops a year, so much that these countries can become 'breadbaskets' for the rest of the world. The food exports benefit people in temperate countries with shorter growing seasons.

- Securing what's in storage: Even after the crop is in, it can be subject to attack by pests. Bugs, moulds, and rodents can harm precious grains. Pesticides used in stored products can prolong the viable life of the produce, prevent huge post-harvest losses from pests and diseases and protect the grain so it is safe to eat.

References

- Pesticides, chemistry: byjus.com, Retrieved 21 July, 2019

- Lide, David R. (2015–2016). "Physical Constants of Organic Compounds". Handbook of Chemistry and Physics (96 ed.). Boca Raton, FL: CRC Press. Pp. 3–122. ISBN 9781482208672

- PHE-ASO-Manual-22042013, PHE-ASO-Manual-22042013: gov.in, Retrieved 22 January, 2019

- "U.S. EPA denies petition to ban pesticide chlorpyrifos", Reuters, March 29, 2017, retrieved March 30, 2017

- ASO-PMD, Recruitments: gov.in, Retrieved 23 February, 2019

- Maddison, Jill E.; Page, Stephen W.; Church, David (2008). Small Animal Clinical Pharmacology. Elsevier Health Sciences. P. 226. ISBN 9780702028588

- Using-acaricide-for-tick-control, pesticides, pests, plant-problems: gardeningknowhow.com, Retrieved 24 March, 2014

- Marina Bjørling-Poulsen; Helle Raun Andersen; Philippe Grandjean. (2008). "Potential developmental neurotoxicity of pesticides used in Europe". Environmental Health. 7: 50. doi:10.1186/1476-069X-7-50. PMC 2577708. PMID 18945337

- Eight-benefits-of-pesticides: croplifeindia.org, Retrieved 26 May, 2019

- Johnson, Reed; Jelmayer, Rogerio (February 15, 2016). "Brazil State Bans Pesticide After Zika Claim". Wall Street Journal. Retrieved February 16, 2016

2

Types of Pesticides

Herbicides, biopesticides and organic pesticides are the three main types of pesticides. Herbicides are the compounds that are used to control weeds and biopesticides are the pesticides derived from naturally occuring materials such as plants, animals and certain minerals. This chapter has been carefully written to provide an easy understanding of these various types of pesticides as well as their applications.

The classification based on the basis of use can be as follows:

- Acaricides: They are the substances that are used to kill mites and ticks, or to disrupt their growth or development and some of the examples are DDT, dicofol, carbofuran, methiocarb, Propoxur, abamectin, milbemectin, flufenoxuron, chlorpyrifos, oxydemeton methyl, Phorate, Phosalone, fenpyroximate, Fipronil, bifenthrin, cyhalothrin, fluvalinate, permethrin, chlorfenapyr.

- Algicide: They are the substances that are used to kill or inhibit algae. Some of the examples are copper sulfate, diuron, isoproturon, isoproturon, oxyfluorfen, simazine.

- Antifeedants: They are the chemicals which prevent an insect or other pest from feeding. Some of the examples are chlordimeforn, fentin and azadirachtin.

- Avicides: They are the chemicals that are used to kill birds. The list include fenthion, strychnine.

- Bactericides: They are the compounds that are isolated from or produced by a microorganism (e.g. a bacterium or a fungus), or a related chemical that is produced artificially which are used to kill or inhibit bacteria in plants or soil. Some of the examples are copper hydroxide, kasugamycin, treptomycin, tetracycline.

- Bird repellents: They are the chemicals which act as the bird repellants and some of the examples are copper oxychloride, diazinon, methiocarb, thiram, ziram.

- Chemosterillant: They are the chemicals that renders an insect infertile and thus prevents it from reproducing. Some insects that mate only once can be controlled or eradicated by releasing huge numbers of sterilised insects, which act as sterilizing substances for the insects.

  All of these act in one of the three ways:

  ◦ They inhibit the production of egg or spam. If it fail then go to the second stages.

- ◦ Cause death of the spam or eggs.

  - ◦ If these steps are failed totally then these bring about lethal mutation on the spam or eggs material and severally damage the genetic material and chromatin material of eggs and spam. This produce zygote, but the off springs will totally lost their reproduction ability.

- Fungicides: They are the chemicals which are used to prevent, cure eradicate the fungi. Some of the examples are cymoxanil , carpropamid, metalaxyl, metalaxyl-M, carboxin, aureofungin, kasugamycin, streptomycin, validamycin, kasugamycin, carbendazim, thiabendazole, thiophanate-methyl, cyproconazole, difenoconazole, flusilazole, tebuconazole, triadimefon, Bordeaux mixture, copper oxychloride, iprodione, captan, ferbam, thiram, ziram, mancozeb, maneb, metiram, propineb, zineb, isoprothiolane, tridemorph, edifenphos, fosetyl-Al, fenarimol, tricyclazole.

- Herbicide softeners: A chemical that protects crops from injury by herbicides, but does not prevent the herbicide from killing weeds. Examples are benoxacor, cloquintocet, cyometrinil, cyprosulfamide.

- Herbicides: They are the substances that are used to kill plants, or to inhibit their growth or development. Some of the examples are alachlor, butachlor, metolachlor, pretilachlor, methabenzthiazuron, pendimethalin, oxyfluorfen, imazethapyr, anilofos, glyphosate, oxadiargyl, oxadiazon, 2,4-D, clodinafop, cyhalofop, quizalofop, Paraquat, atrazine, isoproturon, linuron, metoxuron, chlorimuron, sulfosulfuron.

- Insect attractant: A chemical that lures pests to a trap, thereby removing them from crops, animals or stored products.

- Insect repellents: A chemical that deters an insect from landing on a human or an animal. Some of the examples are Citronella oil, Permethrin.

- Insect Growth regulator: A substance that works by disrupting the growth or development of an insect. Some of the examples are: Diflubenzuron, buprofezin.

- Insecticides: A pesticide that is used to kill insects, or to disrupt their growth or development. Some of the examples are azadirachtin, pyrethrins, carbofuran, carbosulfan, methomyl, buprofezin, diflubenzuron, fenoxycarb, abamectin, emamectin, milbemectin, spinosad, cartap, clothianidin, imidacloprid, thiamethoxam, Acetamiprid, Thiacloprid, DDT, Lindane, Endosulfan, dichlorvos, monocrotophos, phosphamidon, demeton-O-methyl, Ethion, Malathion, phorate, Dimethoate, Phosalone, azinphos-methyl, chlorpyrifos, pirimiphos-methyl, quinalphos, triazophos, cyfluthrin, cyhalothrin, lambda-cyhalothrin, cypermethrin, alpha-cypermethrin, cyphenothrin, deltamethrin, fenpropathrin, esfenvalerate, fluvalinate, imiprothrin, tofenprox, chlorfenapyr, clothianidin thiamethoxam, Thiacloprid, isoprothiolane.

- Mammal repellents: A chemical that deters mammals from approaching or feeding on crops or stored products.

- Mating disrupters: They are the chemicals that interfere with the way that male and female insects locate each other using airborne chemicals (pheromones), thereby preventing them from reproducing.

- Molluscicides: They are the substances used to kill slugs and snails. Some of the examples are copper sulfate, metaldehyde, thiacloprid, thiodicarb.

- Nematicides: They are the chemicals which are used to control Nematicides. Some of the examples are abamectin, benomyl, carbofuran, carbosulfan, methyl bromide, fenamiphos, phosphamidon, chlorpyrifos, dimethoate, phorate, triazophos.

- Plant growth regulators: They are the substances that alters the expected growth, flowering or reproduction rate of plants. Fertilizers and other plant nutrients are excluded from this definition. 2,4-D, α-naphthaleneacetic acid, ethephon, metoxuron, gibberellic acid, chlormequat, paclobutrazol, triacontanol are some of the examples.

- Rodenticides: They are the substances used to kill rats and related animals. Some of the examples are strychnine, bromadiolone, coumachlor, coumatetralyl, warfarin, zinc phosphide, Lindane, aluminium phosphide.

- Synergists: A chemical that enhances the toxicity of a pesticide to a pest, but that is not by itself toxic to the pest. Example: piperonyl butoxide.

- Virucide: An agent having the capacity to destroy or inactivate viruses. Example: Ribavirin.

- Miscellaneous: Aluminium phosphide, sodium cyanide.

- Biologicals: Viruses, bacteria, fungi, and plants Nematodes, insects and other parasites or predators.

Classification on the Basis of the Chemistry

A large number of group of chemicals are available in the list pesticides but we will confine to the pesticides registered.

- Insecticides: The insecticides available can be classified as Organo halogen, Organophosphorous, Carbamates, Pyrethroids, Neonicotinoids, Miscellaneous pesticides, Spinosyns (spinosad) neriestoxin (cartap), Fiproles (or Phenylpyrazoles) (Fipronil), Pyrroles (chlorfenapyr), Quinazolines (fenazaquin), Benzoylureas (diflubenzuron an IGR), Antibiotics (abamectin) etc.

- Fungicides: The fungicides available are aliphatic nitrogen fungicides (dodine), amide fungicides (carpropamid), acylamino acid fungicides (metalaxyl), anilide fungicides (carboxin), antibiotic fungicides (kasugamycin), methoxyacrylate strobilurin fungicides (azoxystrobin), aromatic fungicides (chlorothalonil), carbamate fungicides or benzimidazole fungicides (carbendazim), conazole fungicides (triazoles) (hexaconazole), copper fungicides (COC), dicarboximide fungicides (famoxadone), dichlorophenyl dicarboximide fungicides (iprodione), dinitrophenol fungicides (dinocap), dithiocarbamate fungicides (mancozeb), dithiolane fungicides (isoprothiolane), morpholine fungicides (tridemorph), Sulphur compounds etc.

- Herbicides: The herbicides are anilide herbicides (flufenacet), chloroacetanilide herbicides (butachlor), pyrimidinyloxybenzoic acid herbicides (bispyribac), benzothiazoleherbicides (methabenzthiazuron), dinitroanilineherbicides (pendimethalin), nitrophenyl ether herbicides (oxyfluorfen), halogenated aliphatic herbicides (dalapon), imidazolinone herbicides (imazethapyr), organophosphorus herbicides (anilofos), phenoxyacetic herbicides (2,4-D), aryloxyphenoxypropionic herbicides (clodinafop), quaternary ammonium herbicides (Paraquat), chlorotriazine herbicides (atrazine), triazolone herbicides (carfentrazone), Urea herbicides (methabenzthiazuron), phenylurea herbicides (isoproturon), sulfonylurea herbicides (chlorimuron).

- Rodenticides: Inorganic Rodenticides: (Zinc Phosphide, Aluminium Phosphide,Magnesium Phosphide) coumarin Rodenticides (organic) (bromadiolone, coumachlor, coumatetralyl).

Organochlorine Pesticides

This group consists of, the polychlorinated derivatives of cyclohexane (Lindane), polychlorinated biphenyls (DDT, dicofol) and polychlorinated cyclodiene (Endosulfan).

Properties

Physical

- Solids which possess low volatility.

- Low solubility in water, high solubility in oils, fats, lipids etc.

- Not prone to environmental degradation.

Chemical

- Isomerism is a common phenomenon, ex. Gamma HCH.

- Stable over a wide range of pH.

Toxicity: Possess a high acute toxicity as well as chronic toxicity.

Compound	LD50 (oral) mg/Kg
HCH	1000
dicofol	684-809
lindane	88-91
endosulfan	70-110

By and large these group of chemicals exhibit low selectivity:

- Biological stability: Not rapidly degraded by the enzyme, not rapidly exerted, but get stored in the fatty tissues.

- Behavior in the field: These chemicals are non systemic, act as contact and stomach poisons. Lindane exhibits slight fumigant action. Persist in the environment for long time, results in pesticide residue problem in the environment and bio magnification.

- Effect of OC's in the environment: Insecticides can kill bees, pollination decline and the loss of bees that pollinate plants, and colony collapse disorder (CCD).

A number of the organochlorine pesticides have been banned from most uses worldwide, and globally they are controlled via the Stockholm convention on persistent organic pollutants (POP's).

These include: aldrin, chlordane, DDT, dieldrin, endrin, heptachlor, mirex and toxaphene.

Organo Phosphorous Pesticides

These are the esters of derivatised phosphoric acid, thiophosphoric acid and dithio phosphoric acids which are called phosphates, thiophosphates and dithiophosphates respectively. Some of the examples of each class of pesticides are as follows:

Group	Example
i. Phosphates	Monocrotophos, phosphamidon, DDVP
ii. Thiophosphates or	Methyl parathion, fenitrothion, Phosphorothiates oxy demeton methyl
iii. Dithiophosphates or phosporodithioates	Dimethoate, phorate, Phosalone

Based on the organic moiety attached to the phosphoric acid these grouped can also be classified in to aliphatic, phenyl and heterocyclic derivatives.

Properties

- Physical: These compounds are available as liquids or semi solids and posses significant vapour pressure and comparatively volatile. Some compounds have slight solubility in water (MCP and Phosphamidon are soluble in water). Sunlight brings about modification of the toxicity of these molecules either way.

- Chemical: These compounds which are esters of phosphoric acid are not stable in alkaline pH, but stable over narrow range of pH. Thiophosphates and dithiophosphates undergo molecular rearrangements. Forms isomers with increased toxicity (may take place in storage as well as on application in the field) and under go oxidation to give oxo compounds with increased toxicity. Sulphur atoms in side chain may be oxidised to sulfoxide or sulfones which more toxic than parent molecule.

The organo phosphorous pesticides under go conversion of one pesticide in to another pesticides also takes place. The following are some of the examples.

Trichlorfon	$\Rightarrow$	dichlorvos
Formothion	$\Rightarrow$	dimethoate
Acephate	$\Rightarrow$	methamidophos

- Toxicity: Exhibits acute extreme toxicity to slight toxicity (Phorate1.5 to 3.7 mg / Kg, temephos 8600 mg / Kg). LD50 values may change with the purity of the compound (as the impurities present some times are more toxic than the parent compd) and they do exhibit low chronic toxicity. The undergo rapid conversion in to low fat soluble metabolites which are excreted.

- Biological stability: The OP compounds undergo enzymatic degradation and the metabolites are fat insoluble and easily excreted. Bio magnification is almost absent and chronic toxicity is in significant.

- Behavior in the field: Some of the OP compounds show systemic activity and chemicals like dichlorvos exhibits fumigant action too. Persists for shorter duration and do not pose environmental problem like organo halogen pesticides.

Carbamates

The Carbamates are esters of either carbamic acids or thiocarbamic acids. And the Carbamates may be further subdivided in to three sub-groups as given under:

Group	Example
i. Aryl N methyl carbamate	Carbaryl, Propoxur
ii. Hetero cyclic mono or dimethyl Carbamates	Carbofuran
iii. Carbamoylated oximes	Methomyl
iv. Thiocarbamates	Cartap hydrochloride (neriestoxin group of insecticide)

Properties

- Physical: The organo carbamates are available as non volatile solids. carbaryl, carbofuran have very low water solubility (40-6000ppm) where as Cartaphydrochloride is hygroscopic. And these compounds under go degradation by the environmental factors.

- Chemical: These compounds are unstable in alkaline medium.

- Toxicity: The OC compounds exhibit moderate to extreme toxicity. And they do not display chronic toxicity.

Carbaryl	Moderately toxic
Propoxur and cartap	Highly toxic
Methomyl & carbofuran	Extremely toxic

Biological Stability

The OC compounds undergo enzymatic degradation and rapidly metabolised and excreted. Bio magnification is almost absent and Chronic toxicity is insignificant. Behavior in the field:

Carbaryl & Propoxur	Contact pesticides
Carbofuran, methomyl & Cartap Methomyl is well taken up by leaves Carbofuran is well taken up by the roots (soil applied) and Cartap is taken by both roots and leaves.	Systemic pesticides

Pyrethroids

The living organisms do contain naturally a large number of chemicals some of which give them protection from foreign invasive and many such chemicals have been isolated, identified and evaluated of their biological activity. The flowers of chrysanthemum contain compounds called pyrethrins which are found to have possessed very good pesticidal activity but are found to be less stable in the environment. The pyrethrins are chemically the esters of chrysanthemic acid and pyrethric acid (contains dimethyl cyclopropane group) with alcohols, namely pyrethrolone, cinerolone and jasmolone. Thus there are a total of six esters as shown below.

Ester extract	Acid portion	Alcohol portion	% Content in pyrethrum
Pyrethrin –I	Chrysanthemic acid	Pyrethrolone	35
Cinerin-I	Chrysanthemic acid	Cinerolone	10
Jasmolin-I	Chrysanthemic acid	Jasmolone	5
Pyrethrin –II	Pyrethric acid	Pyrethrolone	32
Cinerin-I	Pyrethric acid	Cinerolone	14
Jasmolin-I	Pyrethric acid	Jasmolone	4

Synthetic Pyrethroids: Allethrin was the first synthetic pyrethroid developed in 1949, followed by resemethrin. However they have failed to contain the desired properties and proved to be highly photo labile. The first photo sable pyrethroid developed was permethrin. This was followed by cypermethrin, deltamethrin, and fenvalerate. Now world over plenty of synthetic Pyrethroids have been synthesized and put to use in plan- protection. The synthetic Pyrethroids contain halogenated derivative of dimethyl cyclopropane carboxylic acid and cyano phenoxy benzyl alcohol. Fenvalerate is an exception with the acid portion being p-chlorophenyl isopropyl acetic acid instead of cyclopropane carboxylic acid. In case of permethrin alcohol portion does not have cyano - group, but it is simply phenoxy benzyl alcohol.

Properties

- Physical: The Pyrethroids are present as volatile and non volatile solids or semisolids. They are insoluble in water.

- Chemical: These compounds present in different isomeric forms Cis isomers are found to be more toxic. Individual isomers also are being marketed (e.g. alphamethrin, deltamethrin). The pyrethroid chemicals are unstable in alkaline medium.

- Toxicity: The toxicity of these chemicals ranges from 80 to 4000mg/Kg body weight and toxicity varies with the ratio of isomers and test animals/species. Toxicity to the insects can be increased by synergists (e.g., Piperonyl butoxide, sesamex).

Neonicotinoids

Neonicotinoids are modified structures from nicotine that have come in to existence with improved bio efficacy.

- Acetamiprid,

- Clothianidin,

- Imidacloprid,

- Thiacloprid,

- Thiamethoxam,

- 2nd generation- thianicotinyl.

Properties

- Physical: These compounds are present as colour less to pale yellow crystals with almost negligible solubility in water with exception of thiamethoxam which has got slight solubility in water.

- Chemical: These compounds are weak bases and are stable in acid conditions. These chemicals under go hydrolysis in alkaline solutions.

- Toxicity: These compounds mostly belong to moderate to high toxic group. They do not have chronic toxicity and not mutagenic and teratogenic. However some of the compounds have been reported to be harmful to honeybees by direct contact but no problems expected when not sprayed into flowering crops.

Miscellaneous Pesticides

The following are new arrivals in to the India from different class of chemical group and with high potential of biological activity. The chemicals like Spinosad and abamectin are biological origin. Many of these chemicals are required to use in very low dose of active ingredient to achieve the pest control and do not bio accumulate.

- Spinosyns (spinosad - Contact and stomach poison) neriestoxin (cartap).

- Fiproles (or Phenylpyrazoles) (Fipronil - Broad spectrum insecticide with contact action).

- Pyrroles (chlorfenapyr - Used against insects and mites stomach and stomach contact action).

- Quinazolines (fenazaquin - Contact acaricide).

- Benzoylureas (diflubenzuron a Contact insect growth regulator IGR).

- Antibiotics (abamectin - Contact and stomach poison-insecticide and acaricide).

HERBICIDES

Herbicides, also commonly known as weedkillers, are chemical substances used to control unwanted plants. Selective herbicides control specific weed species, while leaving the desired crop relatively unharmed, while non-selective herbicides (sometimes called total weedkillers in commercial products) can be used to clear waste ground, industrial and construction sites, railways and railway embankments as they kill all plant material with which they come into contact. Apart from selective/non-selective, other important distinctions include persistence (also known as *residual action*: how long the product stays in place and remains active), *means of uptake* (whether it is absorbed by above-ground foliage only, through the roots, or by other means), and *mechanism of action* (how it works). Historically, products such as common salt and other metal salts were used as herbicides, however these have gradually fallen out of favor and in some countries a number of these are banned due to their persistence in soil, and toxicity and groundwater contamination concerns. Herbicides have also been used in warfare and conflict.

Modern herbicides are often synthetic mimics of natural plant hormones which interfere with growth of the target plants. The term organic herbicide has come to mean herbicides intended for organic farming. Some plants also produce their own natural herbicides, such as the genus *Juglans* (walnuts), or the tree of heaven; such action of natural herbicides, and other related chemical interactions, is called allelopathy. Due to herbicide resistance - a major concern in agriculture - a number of products combine herbicides with different means of action. Integrated pest management may use herbicides alongside other pest control methods.

In the US in 2007, about 83% of all herbicide usage, determined by weight applied, was in agriculture.In 2007, world pesticide expenditures totaled about $39.4 billion; herbicides were about 40% of those sales and constituted the biggest portion, followed by insecticides, fungicides, and other types. Herbicide is also used in forestry, where certain formulations have been found to suppress hardwood varieties in favour of conifers after a clear-cut, as well as pasture systems, and management of areas set aside as wildlife habitat.

Mechanism of Action

Herbicides are often classified according to their site of action, because as a general rule, herbicides within the same site of action class will produce similar symptoms on susceptible plants. Classification based on site of action of herbicide is comparatively better as herbicide resistance management can be handled more properly and effectively. Classification by mechanism of action (MOA) indicates the first enzyme, protein, or biochemical step affected in the plant following application.

List of Mechanisms Found in Modern Herbicides

- ACCase inhibitors: Acetyl coenzyme A carboxylase (ACCase) is part of the first step of lipid synthesis. Thus, ACCase inhibitors affect cell membrane production in the meristems of the grass plant. The ACCases of grasses are sensitive to these herbicides, whereas the ACCases of dicot plants are not.

- ALS inhibitors: The acetolactate synthase (ALS) enzyme (also known as acetohydroxyacid synthase, or AHAS) is the first step in the synthesis of the branched-chain amino acids (valine, leucine, and isoleucine). These herbicides slowly starve affected plants of these amino acids, which eventually leads to inhibition of DNA synthesis. They affect grasses and dicots alike. The ALS inhibitor family includes various sulfonylureas (SUs) (such as Flazasulfuron and Metsulfuron-methyl), imidazolinones (IMIs), triazolopyrimidines (TPs), pyrimidinyl oxybenzoates (POBs), and sulfonylamino carbonyl triazolinones (SCTs). The ALS biological pathway exists only in plants and not animals, thus making the ALS-inhibitors among the safest herbicides.

- EPSPS inhibitors: Enolpyruvylshikimate 3-phosphate synthase enzyme (EPSPS) is used in the synthesis of the amino acids tryptophan, phenylalanine and tyrosine. They affect grasses and dicots alike. Glyphosate (Roundup) is a systemic EPSPS inhibitor inactivated by soil contact.

- The discovery of synthetic auxins inaugurated the era of organic herbicides. They were discovered in the 1940s after long study of the plant growth regulator auxin. Synthetic auxins mimic in some way this plant hormone. They have several points of action on the cell membrane, and are effective in the control of dicot plants. 2,4-D and 2,4,5-T are synthetic auxin herbicides.

- Photosystem II inhibitors reduce electron flow from water to $NADP^+$ at the photochemical step in photosynthesis. They bind to the Qb site on the D1 protein, and prevent quinone from binding to this site. Therefore, this group of compounds causes electrons to accumulate on chlorophyll molecules. As a consequence, oxidation reactions in excess of those normally tolerated by the cell occur, and the plant dies. The triazine herbicides (including atrazine) and urea derivatives (diuron) are photosystem II inhibitors.

- Photosystem I inhibitors steal electrons from the normal pathway through FeS to Fdx to $NADP^+$ leading to direct discharge of electrons on oxygen. As a result, reactive oxygen species are produced and oxidation reactions in excess of those normally tolerated by the cell occur, leading to plant death. Bipyridinium herbicides (such as diquat and paraquat) inhibit the FeS to Fdx step of that chain, while diphenyl ether herbicides (such as nitrofen, nitrofluorfen, and acifluorfen) inhibit the Fdx to $NADP^+$ step.

- HPPD inhibitors inhibit 4-Hydroxyphenylpyruvate dioxygenase, which are involved in tyrosine breakdown. Tyrosine breakdown products are used by plants to make carotenoids, which protect chlorophyll in plants from being destroyed by sunlight. If this happens, the plants turn white due to complete loss of chlorophyll, and the plants die. Mesotrione and sulcotrione are herbicides in this class; a drug, nitisinone, was discovered in the course of developing this class of herbicides.

Herbicide Group (Labeling)

One of the most important methods for preventing, delaying, or managing resistance is to reduce the reliance on a single herbicide mode of action. To do this, farmers must know the mode of action for the herbicides they intend to use, but the relatively complex nature of plant biochemistry makes this difficult to determine. Attempts were made to simplify the understanding of herbicide mode of action by developing a classification system that grouped herbicides by mode of action. Eventually the Herbicide Resistance Action Committee (HRAC) and the Weed Science Society of America (WSSA) developed a classification system. The WSSA and HRAC systems differ in the group designation. Groups in the WSSA and the HRAC systems are designated by numbers and letters, respectively. The goal for adding the "Group" classification and mode of action to the herbicide product label is to provide a simple and practical approach to deliver the information to users. This information will make it easier to develop educational material that is consistent and effective. It should increase user's awareness of herbicide mode of action and provide more accurate recommendations for resistance management. Another goal is to make it easier for users to keep records on which herbicide mode of actions are being used on a particular field from year to year.

Chemical Family

Detailed investigations on chemical structure of the active ingredients of the registered herbicides showed that some moieties (moiety is a part of a molecule that may include either whole functional groups or parts of functional groups as substructures; a functional group has similar chemical properties whenever it occurs in different compounds) have the same mechanisms of action. According to Forouzesh, these moieties have been assigned to the names of chemical families and active ingredients are then classified within the chemical families accordingly. Knowing about herbicide chemical family grouping could serve as a short-term strategy for managing resistance to site of action.

Use and Application

Herbicides being sprayed from the spray arms of a tractor in North Dakota.

Most herbicides are applied as water-based sprays using ground equipment. Ground equipment varies in design, but large areas can be sprayed using self-propelled sprayers equipped with long booms, of 60 to 120 feet (18 to 37 m) with spray nozzles spaced every 20–30 inches (510–760 mm) apart. Towed, handheld, and even horse-drawn sprayers are also used. On large areas, herbicides may also at times be applied aerially using helicopters or airplanes, or through irrigation systems (known as chemigation).

A further method of herbicide application developed around 2010, involves ridding the soil of its active weed seed bank rather than just killing the weed. This can successfully treat annual plants but not perennials. Researchers at the Agricultural Research Service found that the application of herbicides to fields late in the weeds' growing season greatly reduces their seed production, and therefore fewer weeds will return the following season. Because most weeds are annuals, their seeds will only survive in soil for a year or two, so this method will be able to destroy such weeds after a few years of herbicide application.

Weed-wiping may also be used, where a wick wetted with herbicide is suspended from a boom and dragged or rolled across the tops of the taller weed plants. This allows treatment of taller grassland weeds by direct contact without affecting related but desirable shorter plants in the grassland sward beneath. The method has the benefit of avoiding spray drift. In Wales, a scheme offering free weed-wiper hire was launched in 2015 in an effort to reduce the levels of MCPA in water courses.

Misuse and Misapplication

Herbicide volatilisation or spray drift may result in herbicide affecting neighboring fields or plants, particularly in windy conditions. Sometimes, the wrong field or plants may be sprayed due to error.

Use Politically, Militarily and in Conflict

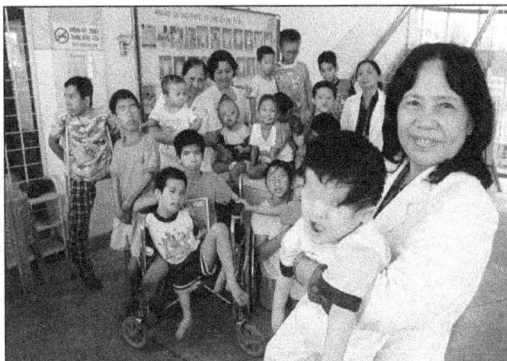

Handicapped children in Vietnam, most of them victims of Agent Orange, 2004.

Although herbicidal warfare use chemical substances, its main purpose is to disrupt agricultural food production and to destroy plants which provide cover or concealment to the enemy.

The use of herbicides as a chemical weapon by the U.S. military during the Vietnam War has left tangible, long-term impacts upon the Vietnamese people that live in Vietnam. For instance, it led to 3 million Vietnamese people suffering health problems, one million birth defects caused directly by exposure to Agent Orange, and 24% of the area of Vietnam being defoliated.

Health and Environmental Effects

Herbicides have widely variable toxicity in addition to acute toxicity arising from ingestion of a significant quantity rapidly, and chronic toxicity arising from environmental and occupational exposure over long periods. Much public suspicion of herbicides revolves around a confusion between valid statements of *acute* toxicity as opposed to equally valid statements of lack of *chronic* toxicity at the recommended levels of usage. For instance, while glyphosate formulations with tallowamine *adjuvants* are acutely toxic, their use was found to be uncorrelated with any health issues like cancer in a massive US Department of Health study on 90,000 members of farmer families for over a period of 23 years. That is, the study shows lack of chronic toxicity, but cannot question the herbicide's acute toxicity.

Some herbicides cause a range of health effects ranging from skin rashes to death. The pathway of attack can arise from intentional or unintentional direct consumption, improper application resulting in the herbicide coming into direct contact with people or wildlife, inhalation of aerial sprays, or food consumption prior to the labelled preharvest interval. Under some conditions, certain herbicides can be transported via leaching or surface runoff to contaminate groundwater or distant surface water sources. Generally, the conditions that promote herbicide transport include intense storm events (particularly shortly after application) and soils with limited capacity to adsorb or retain the herbicides. Herbicide properties that increase likelihood of transport include persistence (resistance to degradation) and high water solubility.

Phenoxy herbicides are often contaminated with dioxins such as TCDD; research has suggested such contamination results in a small rise in cancer risk after occupational exposure to these herbicides. Triazine exposure has been implicated in a likely relationship to increased risk of breast cancer, although a causal relationship remains unclear.

Herbicide manufacturers have at times made false or misleading claims about the safety of their products. Chemical manufacturer Monsanto Company agreed to change its advertising after pressure from New York attorney general Dennis Vacco; Vacco complained about misleading claims that its spray-on glyphosate-based herbicides, including Roundup, were safer than table salt and "practically non-toxic" to mammals, birds, and fish (though proof that this was ever said is hard to find). Roundup is toxic and has resulted in death after being ingested in quantities ranging from 85 to 200 ml, although it has also been ingested in quantities as large as 500 ml with only mild or moderate symptoms. The manufacturer of Tordon 101 (Dow AgroSciences, owned by the Dow Chemical Company) has claimed Tordon 101 has no effects on animals and insects, in spite of evidence of strong carcinogenic activity of the active ingredient Picloram in studies on rats.

The risk of Parkinson's disease has been shown to increase with occupational exposure to herbicides and pesticides. The herbicide paraquat is suspected to be one such factor.

All commercially sold, organic and nonorganic herbicides must be extensively tested prior to approval for sale and labeling by the Environmental Protection Agency. However, because of the large number of herbicides in use, concern regarding health effects is significant. In addition to health effects caused by herbicides themselves, commercial herbicide mixtures often contain other chemicals, including inactive ingredients, which have negative impacts on human health.

Ecological Effects

Commercial herbicide use generally has negative impacts on bird populations, although the impacts are highly variable and often require field studies to predict accurately. Laboratory studies have at times overestimated negative impacts on birds due to toxicity, predicting serious problems that were not observed in the field. Most observed effects are due not to toxicity, but to habitat changes and the decreases in abundance of species on which birds rely for food or shelter. Herbicide use in silviculture, used to favor certain types of growth following clearcutting, can cause significant drops in bird populations. Even when herbicides which have low toxicity to birds are used, they decrease the abundance of many types of vegetation on which the birds rely. Herbicide use in agriculture in Britain has been linked to a decline in seed-eating bird species which rely on the weeds killed by the herbicides. Heavy use of herbicides in neotropical agricultural areas has been one of many factors implicated in limiting the usefulness of such agricultural land for wintering migratory birds.

Frog populations may be affected negatively by the use of herbicides as well. While some studies have shown that atrazine may be a teratogen, causing demasculinization in male frogs, the U.S. Environmental Protection Agency (EPA) and its independent Scientific Advisory Panel (SAP) examined all available studies on this topic and concluded that "atrazine does not adversely affect amphibian gonadal development based on a review of laboratory and field studies."

Scientific Uncertainty of Full Extent of Herbicide Effects

The health and environmental effects of many herbicides is unknown, and even the scientific community often disagrees on the risk. For example, a 1995 panel of 13 scientists reviewing studies on the carcinogenicity of 2,4-D had divided opinions on the likelihood 2,4-D causes cancer in humans. As of 1992, studies on phenoxy herbicides were too few to accurately assess the risk of many types of cancer from these herbicides, even though evidence was stronger that exposure to these herbicides is associated with increased risk of soft tissue sarcoma and non-Hodgkin lymphoma. Furthermore, there is some suggestion that herbicides can play a role in sex reversal of certain organisms that experience temperature-dependent sex determination, which could theoretically alter sex ratios.

Resistance

Weed resistance to herbicides has become a major concern in crop production worldwide. Resistance to herbicides is often attributed to lack of rotational programmes of herbicides and to continuous applications of herbicides with the same sites of action. Thus, a true understanding of the sites of action of herbicides is essential for strategic planning of herbicide-based weed control.

Plants have developed resistance to atrazine and to ALS-inhibitors, and more recently, to glyphosate herbicides. Marestail is one weed that has developed glyphosate resistance. Glyphosate-resistant weeds are present in the vast majority of soybean, cotton and corn farms in some U.S. states. Weeds that can resist multiple other herbicides are spreading. Few new herbicides are near commercialization, and none with a molecular mode of action for which there is no resistance. Because most herbicides could not kill all weeds, farmers rotated crops and herbicides to stop resistant weeds. During its initial years, glyphosate was not subject to resistance and allowed farmers to reduce the use of rotation.

A family of weeds that includes waterhemp (Amaranthus rudis) is the largest concern. A 2008-9 survey of 144 populations of waterhemp in 41 Missouri counties revealed glyphosate resistance in 69%. Weeds from some 500 sites throughout Iowa in 2011 and 2012 revealed glyphosate resistance in approximately 64% of waterhemp samples. The use of other killers to target "residual" weeds has become common, and may be sufficient to have stopped the spread of resistance. From 2005 through 2010 researchers discovered 13 different weed species that had developed resistance to glyphosate. But since then only two more have been discovered. Weeds resistant to multiple herbicides with completely different biological action modes are on the rise. In Missouri, 43% of samples were resistant to two different herbicides; 6% resisted three; and 0.5% resisted four. In Iowa 89% of waterhemp samples resist two or more herbicides, 25% resist three, and 10% resist five.

For southern cotton, herbicide costs has climbed from between $50 and $75 per hectare a few years ago to about $370 per hectare in 2013. Resistance is contributing to a massive shift away from growing cotton; over the past few years, the area planted with cotton has declined by 70% in Arkansas and by 60% in Tennessee. For soybeans in Illinois, costs have risen from about $25 to $160 per hectare.

Dow, Bayer CropScience, Syngenta and Monsanto are all developing seed varieties resistant to herbicides other than glyphosate, which will make it easier for farmers to use alternative weed killers. Even though weeds have already evolved some resistance to those herbicides, Powles says the new seed-and-herbicide combos should work well if used with proper rotation.

Biochemistry of resistance

Resistance to herbicides can be based on one of the following biochemical mechanisms:

- Target-site resistance: This is due to a reduced (or even lost) ability of the herbicide to bind to its target protein. The effect usually relates to an enzyme with a crucial function in a metabolic pathway, or to a component of an electron-transport system. Target-site resistance may also be caused by an overexpression of the target enzyme (via gene amplification or changes in a gene promoter).

- Non-target-site resistance: This is caused by mechanisms that reduce the amount of herbicidal active compound reaching the target site. One important mechanism is an enhanced metabolic detoxification of the herbicide in the weed, which leads to insufficient amounts of the active substance reaching the target site. A reduced uptake and translocation, or sequestration of the herbicide, may also result in an insufficient herbicide transport to the target site.

- Cross-resistance: In this case, a single resistance mechanism causes resistance to several herbicides. The term target-site cross-resistance is used when the herbicides bind to the same target site, whereas non-target-site cross-resistance is due to a single non-target-site mechanism (e.g., enhanced metabolic detoxification) that entails resistance across herbicides with different sites of action.

- Multiple resistance: In this situation, two or more resistance mechanisms are present within individual plants, or within a plant population.

Resistance Management

Worldwide Experience has been that farmers tend to do little to prevent herbicide resistance developing, and only take action when it is a problem on their own farm or neighbor's. Careful observation is important so that any reduction in herbicide efficacy can be detected. This may indicate evolving resistance. It is vital that resistance is detected at an early stage as if it becomes an acute, whole-farm problem, options are more limited and greater expense is almost inevitable. Table lists factors which enable the risk of resistance to be assessed. An essential pre-requisite for confirmation of resistance is a good diagnostic test. Ideally this should be rapid, accurate, cheap and accessible. Many diagnostic tests have been developed, including glasshouse pot assays, petri dish assays and chlorophyll fluorescence. A key component of such tests is that the response of the suspect population to a herbicide can be compared with that of known susceptible and resistant standards under controlled conditions. Most cases of herbicide resistance are a consequence of the repeated use of herbicides, often in association with crop monoculture and reduced cultivation practices. It is necessary, therefore, to modify these practices in order to prevent or delay the onset of resistance or to control existing resistant populations. A key objective should be the reduction in selection pressure. An integrated weed management (IWM) approach is required, in which as many tactics as possible are used to combat weeds. In this way, less reliance is placed on herbicides and so selection pressure should be reduced.

Optimising herbicide input to the economic threshold level should avoid the unnecessary use of herbicides and reduce selection pressure. Herbicides should be used to their greatest potential by ensuring that the timing, dose, application method, soil and climatic conditions are optimal for good activity. In the UK, partially resistant grass weeds such as *Alopecurus myosuroides* (blackgrass) and *Avena* spp. (wild oat) can often be controlled adequately when herbicides are applied at the 2-3 leaf stage, whereas later applications at the 2-3 tiller stage can fail badly. Patch spraying, or applying herbicide to only the badly infested areas of fields, is another means of reducing total herbicide use.

Table: Agronomic factors influencing the risk of herbicide resistance development.

Factor	Low risk	High risk
Cropping system	Good rotation	Crop monoculture
Cultivation system	Annual ploughing	Continuous minimum tillage
Weed control	Cultural only	Herbicide only
Herbicide use	Many modes of action	Single modes of action
Control in previous years	Excellent	Poor
Weed infestation	Low	High
Resistance in vicinity	Unknown	Common

Approaches to Treating Resistant Weeds

Alternative Herbicides

When resistance is first suspected or confirmed, the efficacy of alternatives is likely to be the first consideration. The use of alternative herbicides which remain effective on resistant populations can be a successful strategy, at least in the short term. The effectiveness of alternative herbicides will be highly dependent on the extent of cross-resistance. If there is resistance to a single group of herbicides, then the use of herbicides from other groups may provide a simple and effective solution, at least in the short term. For example, many triazine-resistant weeds have been readily controlled by the use of alternative herbicides such as dicamba or glyphosate. If resistance extends to more than one herbicide group, then choices are more limited. It should not be assumed that resistance will automatically extend to all herbicides with the same mode of action, although it is wise to assume this until proved otherwise. In many weeds the degree of cross-resistance between the five groups of ALS inhibitors varies considerably. Much will depend on the resistance mechanisms present, and it should not be assumed that these will necessarily be the same in different populations of the same species. These differences are due, at least in part, to the existence of different mutations conferring target site resistance. Consequently, selection for different mutations may result in different patterns of cross-resistance. Enhanced metabolism can affect even closely related herbicides to differing degrees. For example, populations of *Alopecurus myosuroides* (blackgrass) with an enhanced metabolism mechanism show resistance to pendimethalin but not to trifluralin, despite both being dinitroanilines. This is due to differences in the vulnerability of these two herbicides to oxidative metabolism. Consequently, care is needed when trying to predict the efficacy of alternative herbicides.

Mixtures and Sequences

The use of two or more herbicides which have differing modes of action can reduce the selection for resistant genotypes. Ideally, each component in a mixture should:

- Be active at different target sites.

- Have a high level of efficacy.

- Be detoxified by different biochemical pathways.

- Have similar persistence in the soil (if it is a residual herbicide).

- Exert negative cross-resistance.

- Synergise the activity of the other component.

No mixture is likely to have all these attributes, but the first two listed are the most important. There is a risk that mixtures will select for resistance to both components in the longer term. One practical advantage of sequences of two herbicides compared with mixtures is that a better appraisal of the efficacy of each herbicide component is possible, provided that sufficient time elapses between each application. A disadvantage with sequences is that two separate applications have to be made and it is possible that the later application will be less effective on weeds surviving the first application. If these are resistant, then the second herbicide in the sequence may increase selection for resistant individuals by killing the susceptible plants which were damaged but not killed by the first application, but allowing the larger, less affected, resistant plants to survive. This has been cited as one reason why ALS-resistant *Stellaria media* has evolved in Scotland recently, despite the regular use of a sequence incorporating mecoprop, a herbicide with a different mode of action.

Herbicide Rotations

Rotation of herbicides from different chemical groups in successive years should reduce selection for resistance. This is a key element in most resistance prevention programmes. The value of this approach depends on the extent of cross-resistance, and whether multiple resistance occurs owing to the presence of several different resistance mechanisms. A practical problem can be the lack of awareness by farmers of the different groups of herbicides that exist. In Australia a scheme has been introduced in which identifying letters are included on the product label as a means of enabling farmers to distinguish products with different modes of action.

Farming Practices and Resistance

Herbicide resistance became a critical problem in Australian agriculture, after many Australian sheep farmers began to exclusively grow wheat in their pastures in the 1970s. Introduced varieties of ryegrass, while good for grazing sheep, compete intensely with wheat. Ryegrasses produce so many seeds that, if left unchecked, they can completely choke a field. Herbicides provided excellent control, while reducing soil disrupting because of less need to plough. Within little more than a decade, ryegrass and other weeds began to develop resistance. In response Australian farmers changed methods. By 1983, patches of ryegrass had become immune to Hoegrass, a family of herbicides that inhibit an enzyme called acetyl coenzyme A carboxylase.

Ryegrass populations were large, and had substantial genetic diversity, because farmers had planted many varieties. Ryegrass is cross-pollinated by wind, so genes shuffle frequently. To control its distribution farmers sprayed inexpensive Hoegrass, creating selection pressure. In addition, farmers sometimes diluted the herbicide in order to save money, which allowed some plants to survive application. When resistance appeared farmers turned to a group of herbicides that block acetolactate synthase. Once again, ryegrass in Australia evolved a kind of "cross-resistance" that allowed it to rapidly break down a variety of herbicides. Four classes of herbicides become ineffective within a few years. In 2013 only two herbicide classes, called Photosystem II and long-chain fatty acid inhibitors, were effective against ryegrass.

List of Common Herbicides

Synthetic Herbicides

- 2,4-D is a broadleaf herbicide in the phenoxy group used in turf and no-till field crop production. Now, it is mainly used in a blend with other herbicides to allow lower rates of herbicides to be used; it is the most widely used herbicide in the world, and third most commonly used in the United States. It is an example of synthetic auxin (plant hormone).

- Aminopyralid is a broadleaf herbicide in the pyridine group, used to control weeds on grassland, such as docks, thistles and nettles. It is notorious for its ability to persist in compost.

- Atrazine, a triazine herbicide, is used in corn and sorghum for control of broadleaf weeds and grasses. Still used because of its low cost and because it works well on a broad spectrum of weeds common in the US corn belt, atrazine is commonly used with other herbicides to reduce the overall rate of atrazine and to lower the potential for groundwater contamination; it is a photosystem II inhibitor.

- Clopyralid is a broadleaf herbicide in the pyridine group, used mainly in turf, rangeland, and for control of noxious thistles. Notorious for its ability to persist in compost, it is another example of synthetic auxin.

- Dicamba, a postemergent broadleaf herbicide with some soil activity, is used on turf and field corn. It is another example of a synthetic auxin.

- Glufosinate ammonium, a broad-spectrum contact herbicide, is used to control weeds after the crop emerges or for total vegetation control on land not used for cultivation.

- Fluazifop (Fuselade Forte), a post emergence, foliar absorbed, translocated grass-selective herbicide with little residual action. It is used on a very wide range of broad leaved crops for control of annual and perennial grasses.

- Fluroxypyr, a systemic, selective herbicide, is used for the control of broad-leaved weeds in small grain cereals, maize, pastures, rangeland and turf. It is a synthetic auxin. In cereal growing, fluroxypyr's key importance is control of cleavers, Galium aparine. Other key broadleaf weeds are also controlled.

- Glyphosate, a systemic nonselective herbicide, is used in no-till burndown and for weed control in crops genetically modified to resist its effects. It is an example of an EPSPs inhibitor.

- Imazapyr a nonselective herbicide, is used for the control of a broad range of weeds, including terrestrial annual and perennial grasses and broadleaf herbs, woody species, and riparian and emergent aquatic species.

- Imazapic, a selective herbicide for both the pre- and postemergent control of some annual and perennial grasses and some broadleaf weeds, kills plants by inhibiting the production of branched chain amino acids (valine, leucine, and isoleucine), which are necessary for protein synthesis and cell growth.

- Imazamox, an imidazolinone manufactured by BASF for postemergence application that is an acetolactate synthase (ALS) inhibitor. Sold under trade names Raptor, Beyond, and Clearcast.

- Linuron is a nonselective herbicide used in the control of grasses and broadleaf weeds. It works by inhibiting photosynthesis.

- MCPA (2-methyl-4-chlorophenoxyacetic acid) is a phenoxy herbicide selective for broadleaf plants and widely used in cereals and pasture.

- Metolachlor is a pre-emergent herbicide widely used for control of annual grasses in corn and sorghum; it has displaced some of the atrazine in these uses.

- Paraquat is a nonselective contact herbicide used for no-till burndown and in aerial destruction of marijuana and coca plantings. It is more acutely toxic to people than any other herbicide in widespread commercial use.

- Pendimethalin, a pre-emergent herbicide, is widely used to control annual grasses and some broad-leaf weeds in a wide range of crops, including corn, soybeans, wheat, cotton, many tree and vine crops, and many turfgrass species.

- Picloram, a pyridine herbicide, mainly is used to control unwanted trees in pastures and edges of fields. It is another synthetic auxin.

- Sodium chlorate (disused/banned in some countries), a nonselective herbicide, is considered phytotoxic to all green plant parts. It can also kill through root absorption.

- Triclopyr, a systemic, foliar herbicide in the pyridine group, is used to control broadleaf weeds while leaving grasses and conifers unaffected.

- Several sulfonylureas, including Flazasulfuron and Metsulfuron-methyl, which act as ALS inhibitors and in some cases are taken up from the soil via the roots.

Organic Herbicides

Recently, the term "organic" has come to imply products used in organic farming. Under this definition, an organic herbicide is one that can be used in a farming enterprise that has been classified as organic. Depending on the application, they may be less effective than synthetic herbicides and are generally used along with cultural and mechanical weed control practices.

Homemade organic herbicides include:

- Corn gluten meal (CGM) is a natural pre-emergence weed control used in turfgrass, which reduces germination of many broadleaf and grass weeds.

- Vinegar is effective for 5–20% solutions of acetic acid, with higher concentrations most effective, but it mainly destroys surface growth, so respraying to treat regrowth is needed. Resistant plants generally succumb when weakened by respraying.

- Steam has been applied commercially, but is now considered uneconomical and inadequate. It controls surface growth but not underground growth and so respraying to treat regrowth of perennials is needed.

- Flame is considered more effective than steam, but suffers from the same difficulties.

- D-limonene (citrus oil) is a natural degreasing agent that strips the waxy skin or cuticle from weeds, causing dehydration and ultimately death.

- Saltwater or salt applied in appropriate strengths to the rootzone will kill most plants.

- 2,4,5-Trichlorophenoxyacetic acid (2,4,5-T) was a widely used broadleaf herbicide until being phased out starting in the late 1970s. While 2,4,5-T itself is of only moderate toxicity, the manufacturing process for 2,4,5-T contaminates this chemical with trace amounts of 2,3,7,8-tetrachlorodibenzo-p-dioxin (TCDD). TCDD is extremely toxic to humans. With proper temperature control during production of 2,4,5-T, TCDD levels can be held to about .005 ppm. Before the TCDD risk was well understood, early production facilities lacked proper temperature controls. Individual batches tested later were found to have as much as 60 ppm of TCDD. 2,4,5-T was withdrawn from use in the USA in 1983, at a time of heightened public sensitivity about chemical hazards in the environment. Public concern about dioxins was high, and production and use of other (non-herbicide) chemicals potentially containing TCDD contamination was also withdrawn. These included pentachlorophenol (a wood preservative) and PCBs (mainly used as stabilizing agents in transformer oil). Some feel that the 2,4,5-T withdrawal was not based on sound science. 2,4,5-T has since largely been replaced by dicamba and triclopyr.

- Agent Orange was a herbicide blend used by the British military during the Malayan Emergency and the U.S. military during the Vietnam War between January 1965 and April 1970 as a defoliant. It was a 50/50 mixture of the *n*-butyl esters of 2,4,5-T and 2,4-D. Because of TCDD contamination in the 2,4,5-T component, it has been blamed for serious illnesses in many people who were exposed to it.

- Diesel, and other heavy oil derivatives, are known to be informally used at times, but are usually banned for this purpose.

BIOPESTICIDES

Biopesticide is a contraction of 'biological pesticides' which includes several types of pest management intervention: through predatory, parasitic, or chemical relationships. The term has been associated historically with biological control – and by implication – the manipulation of living organisms. Regulatory positions can be influenced by public perceptions, thus:

- In the EU, biopesticides have been defined as "a form of pesticide based on micro-organisms or natural products".

- The US EPA states that they "include naturally occurring substances that control pests (biochemical pesticides), microorganisms that control pests (microbial pesticides), and pesticidal substances produced by plants containing added genetic material (plant-incorporated protectants) or PIPs".

They are obtained from organisms including plants, bacteria and other microbes, fungi, nematodes, etc. They are often important components of integrated pest management (IPM) programmes, and have received much practical attention as substitutes to synthetic chemical plant protection products (PPPs).

Types

Biopesticides can be classified into these classes:

- Microbial pesticides which consist of bacteria, entomopathogenic fungi or viruses (and sometimes includes the metabolites that bacteria or fungi produce). Entomopathogenic nematodes are also often classed as microbial pesticides, even though they are multi-cellular.

- Bio-derived chemicals. Four groups are in commercial use: pyrethrum, rotenone, neem oil, and various essential oils are naturally occurring substances that control (or monitor in the case of pheromones) pests and microbial diseases.

- Plant-incorporated protectants (PIPs) have genetic material from other species incorporated into their genetic material (i.e. GM crops). Their use is controversial, especially in many European countries.

- RNAi pesticides, some of which are topical and some of which are absorbed by the crop.

Biopesticides have usually no known function in photosynthesis, growth or other basic aspects of plant physiology. Instead, they are active against biological pests. Many chemical compounds have been identified that are produced by plants to protect them from pests so they are called antifeedants. These materials are biodegradable and renewable alternatives, which can be economical for practical use. Organic farming systems embraces this approach to pest control.

RNA

RNA interference is under study for possible use as a spray-on insecticide by multiple companies, including Monsanto, Syngenta, and Bayer. Such sprays do not modify the genome of the target plant. The RNA could be modified to maintain its effectiveness as target species evolve tolerance to the original. RNA is a relatively fragile molecule that generally degrades within days or weeks of application. Monsanto estimated costs to be on the order of $5/acre.

RNAi has been used to target weeds that tolerate Monsanto's Roundup herbicide. RNAi mixed with a silicone surfactant that let the RNA molecules enter air-exchange holes in the plant's surface that disrupted the gene for tolerance, affecting it long enough to let the herbicide work. This strategy would allow the continued use of glyphosate-based herbicides, but would not per se assist a herbicide rotation strategy that relied on alternating Roundup with others.

They can be made with enough precision to kill some insect species, while not harming others. Monsanto is also developing an RNA spray to kill potato beetles One challenge is to make it linger on the plant for a week, even if it's raining. The Potato beetle has become resistant to more than 60 conventional insecticides.

Monsanto lobbied the U.S. EPA to exempt RNAi pesticide products from any specific regulations (beyond those that apply to all pesticides) and be exempted from rodent toxicity, allergenicity and residual environmental testing. In 2014 an EPA advisory group found little evidence of a risk to people from eating RNA.

However, in 2012, the Australian Safe Food Foundation alleged that the RNA trigger designed to change wheat's starch content might interfere with the gene for a human liver enzyme. Supporters countered that RNA does not appear to make it past human saliva or stomach acids. The US National Honey Bee Advisory Board told EPA that using RNAi would put natural systems at "the epitome of risk". The beekeepers cautioned that pollinators could be hurt by unintended effects and that the genomes of many insects are still unknown. Other unassessed risks include ecological (given the need for sustained presence for herbicide and other applications) and the possible for RNA drift across species boundaries.

Monsanto has invested in multiple companies for their RNA expertise, including Beeologics (for RNA that kills a parasitic mite that infests hives and for manufacturing technology) and Preceres (nanoparticle lipidoid coatings) and licensed technology from Alnylam and Tekmira. In 2012 Syngenta acquired Devgen, a European RNA partner. Startup Forrest Innovations is investigating RNAi as a solution to citrus greening disease that in 2014 caused 22 percent of oranges in Florida to fall off the trees.

Examples:

Bacillus thuringiensis, a bacterial disease of Lepidoptera, Coleoptera and Diptera, is a well-known insecticide example. The toxin from B. thuringiensis (Bt toxin) has been incorporated directly into plants through the use of genetic engineering. The use of Bt Toxin is particularly controversial. Its manufacturers claim it has little effect on other organisms, and is more environmentally friendly than synthetic pesticides.

Other microbial control agents include products based on:

- Entomopathogenic fungi (e.g. Beauveria bassiana, Isaria fumosorosea, Lecanicillium and Metarhizium spp.).

- Plant disease control agents, include Trichoderma spp. and Ampelomyces quisqualis (a hyper-parasite of grape powdery mildew); Bacillus subtilis is also used to control plant pathogens.

- Beneficial nematodes attacking insect (e.g., Steinernema feltiae) or slug (e.g. Phasmarhabditis hermaphrodita) pests.

- Entomopathogenic viruses (e.g., Cydia pomonella granulovirus).

- Weeds and rodents have also been controlled with microbial agents.

Various naturally occurring materials, including fungal and plant extracts, have been described as biopesticides. Products in this category include:

- Insect pheromones and other semiochemicals.

- Fermentation products such as Spinosad (a macro-cyclic lactone).

- Chitosan: A plant in the presence of this product will naturally induce systemic resistance (ISR) to allow the plant to defend itself against disease, pathogens and pests.

- Biopesticides may include natural plant-derived products, which include alkaloids, terpenoids, phenolics and other secondary chemicals. Certain vegetable oils such as canola oil are known to have pesticidal properties. Products based on extracts of plants such as garlic have now been registered in the EU and elsewhere.

Applications

Biopesticides are biological or biologically-derived agents, that are usually applied in a manner similar to chemical pesticides, but achieve pest management in an environmentally friendly way. With all pest management products, but especially microbial agents, effective control requires appropriate formulation and application.

Biopesticides for use against crop diseases have already established themselves on a variety of crops. For example, biopesticides already play an important role in controlling downy mildew diseases. Their benefits include: a 0-Day Pre-Harvest Interval, the ability to use under moderate to severe disease pressure, and the ability to use as a tank mix or in a rotational program with other registered fungicides. Because some market studies estimate that as much as 20% of global fungicide sales are directed at downy mildew diseases, the integration of biofungicides into grape production has substantial benefits in terms of extending the useful life of other fungicides, especially those in the reduced-risk category.

A major growth area for biopesticides is in the area of seed treatments and soil amendments. Fungicidal and biofungicidal seed treatments are used to control soil borne fungal pathogens that cause seed rots, damping-off, root rot and seedling blights. They can also be used to control internal seed–borne fungal pathogens as well as fungal pathogens that are on the surface of the seed. Many biofungicidal products also show capacities to stimulate plant host defence and other physiological processes that can make treated crops more resistant to a variety of biotic and abiotic stresses.

Disadvantages

- High specificity: Which may require an exact identification of the pest/pathogen and the use of multiple products to be used; although this can also be an advantage in that the biopesticide is less likely to harm species other than the target.

- Often slow speed of action (thus making them unsuitable if a pest outbreak is an immediate threat to a crop).

- Often variable efficacy due to the influences of various biotic and abiotic factors (since some biopesticides are living organisms, which bring about pest/pathogen control by multiplying within or nearby the target pest/pathogen).

- Living organisms evolve and increase their resistance to biological, chemical, physical or any other form of control. If the target population is not exterminated or rendered incapable of reproduction, the surviving population can acquire a tolerance of whatever pressures are brought to bear, resulting in an evolutionary arms race.

- Unintended consequences: Studies have found broad spectrum biopesticides have lethal and nonlethal risks for non-target native pollinators such as Melipona quadrifasciata in Brazil.

ORGANIC PESTICIDES

Organic pesticides are often considered safer than non-organic pesticides for the environment, people and animals.

Identification

Organic pesticides are made from naturally occurring ingredients. Non-organic pesticides are created synthetically.

Organic pesticides are used by professional flower and vegetable garden businesses that want or want to keep their organic certification and by home gardeners who want a natural alternative to non-organic pesticides.

Considerations

"Organic" does not mean non-toxic. It is important to read and follow the directions for each organic pesticide. A drawback to organic pesticides is that they usually have to be reapplied numerous times, possibly making a larger impact on the environment than a conventional, non-organic pesticide that is used less often.

Types

Organic pesticides are available in a wide variety of applications. Insecticidal soap, powdered bacteria such as Bacillus thuringiensis (Bt) and pyrethrins, which are chemicals derived from plants, are organic products commonly found in most garden stores. These pesticides are used to kill insects on contact and to keep them from reproducing in gardens.

Warnings

Organic pesticides are a good idea for any gardener if used properly. Although the pesticides are labeled "organic," proper safety gear and application rate must be followed for best results.

Neem

Ancient Indians highly revered neem oil as a powerful, all-natural plant for warding off pests. Neem juice is even one the most powerful natural pesticides on the planet, holding over 50 natural insecticides. You can use this extremely bitter tree leaf to make a natural pesticidal spray.

To make neem oil spray, add half an ounce of high-quality organic neem oil and half a teaspoon of a mild organic liquid soap to two quarts of warm water. Stir slowly. Add to a spray bottle and use immediately.

Himalayan crystal salt.

Salt Spray

For treating plants infested with spider mites, mix two tablespoons of Himalayan Crystal Salt into one gallon of warm water and spray on infected areas.

Mineral Oil

Mix 10-30 ml of high-grade oil with one liter of water. Stir and add to spray bottle. This organic pesticide works well for dehydrating insects and their eggs.

Citrus Oil and Cayenne Pepper

This organic pesticide works well on ants. Mix ten drops of citrus essential oil with one teaspoon cayenne pepper and 1 cup of warm water. Shake well and spray on the affected areas.

Soap, Orange Citrus Oil and Water

To make this natural pesticide, simply mix three tablespoons of liquid Organic Castile soap with 1 ounce of Orange oil to one gallon of water. Shake well. This is an especially effective treatment against slugs and can be sprayed directly on ants and roaches.

Eucalyptus Oil

A great natural pesticide for flies, bees, and wasps. Simply sprinkle a few drops of eucalyptus oil where the insects are found. They will all be gone before you know it.

Onion and Garlic Spray

Mince one organic clove of garlic and one medium-sized organic onion. Add to a quart of water. Wait one hour and then add one teaspoon of cayenne pepper and one tablespoon of liquid soap to the mix. This organic spray will hold its potency for one week if stored in the refrigerator.

Chrysanthemum Flower Tea

These flowers hold a powerful plant chemical component called pyrethrum. This substance invades the nervous system of insects, rendering them immobile. You can make your own spray by boiling 100 grams of dried flowers into 1 liter of water. Boil dried flowers in water for twenty minutes. Strain, cool, and pour into a spray bottle. Can be stored for up to two months. You can also add some organic neem oil to enhance the effectiveness.

Tobacco Spray

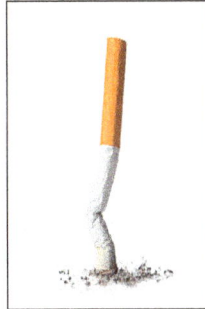

Tobacco.

Just as tobacco is hazardous to humans, tobacco spray was once a commonly used pesticide for killing pests, caterpillars, and aphids. Mix one cup of organic tobacco (preferably a brand that is organic and all-natural) into one gallon of water. Allow the mixture to set overnight. After 24-hours, the mix should have a light brown color. If it is very dark, add more water. This mix can be used on most plants, except those in the solanaceous family (tomatoes, peppers, eggplants, etc.)

Chile Pepper and Diatomaceous Earth

Grind two handfuls of dry chiles into a fine powder and mix with one cup of diatomaceous earth. Add to two liters of water and let sit overnight. Shake well before applying.

References

- Environmental Protection Agency: Atrazine Updates.Current as of January 2013. Retrieved August 24, 2013

- Robbins, Paul (2007-08-27). Encyclopedia of environment and society. Robbins, Paul, 1967-, Sage Publications. Thousand Oaks. P. 862. ISBN 9781452265582

- Definition-of-organic-pesticides-13406429: hunker.com, Retrieved 28 July, 2019

- Murray B. Isman "Botanical Insecticides, Deterrents, And Repellents In Modern Agriculture And An Increasingly Regulated World" Annual Review Of Entomology Volume 51, pp. 45-66. Doi:10.1146/annurev.ento.51.110104.151146

- Organic-pesticides, natural-health: natural-health, Retrieved 29 August, 2019

- Copping, Leonard G. (2009). The Manual of Biocontrol Agents: A World Compendium. BCPC. ISBN 978-1-901396-17-1

3

Insecticides and its Types

The agents of chemical and biological origin which control or kill insects are referred to as insecticides. Some of the common types of insecticides are organochlorines, organophosphates, organosulfur formamidines, dinitrophenols, fiproles, etc. All these types of insecticides have been carefully analyzed in this chapter.

INSECTICIDE

Insecticide is any toxic substance that is used to kill insects. Such substances are used primarily to control pests that infest cultivated plants or to eliminate disease-carrying insects in specific areas.

Insecticides can be classified in any of several ways, on the basis of their chemistry, their toxicological action, or their mode of penetration. In the latter scheme, they are classified according to whether they take effect upon ingestion (stomach poisons), inhalation (fumigants), or upon penetration of the body covering (contact poisons). Most synthetic insecticides penetrate by all three of these pathways, however, and hence are better distinguished from each other by their basic chemistry. Besides the synthetics, some organic compounds occurring naturally in plants are useful insecticides, as are some inorganic compounds; some of these are permitted in organic farming applications. Most insecticides are sprayed or dusted onto plants and other surfaces traversed or fed upon by insects.

Modes of Penetration

Stomach poisons are toxic only if ingested through the mouth and are most useful against those insects that have biting or chewing mouth parts, such as caterpillars, beetles, and grasshoppers. The chief stomach poisons are the arsenicals—e.g., Paris green (copper acetoarsenite), lead arsenate, and calcium arsenate; and the fluorine compounds, among them sodium fluoride and cryolite. They are applied as sprays or dusts onto the leaves and stems of plants eaten by the target insects. Stomach poisons have gradually been replaced by synthetic insecticides, which are less dangerous to humans and other mammals.

Contact poisons penetrate the skin of the pest and are used against those arthropods, such as aphids, that pierce the surface of a plant and suck out the juices. The contact insecticides can be divided into two main groups: naturally occurring compounds and synthetic organic ones. The naturally occurring contact insecticides include nicotine, developed from tobacco; pyrethrum, obtained from flowers of Chrysanthemum cinerariaefolium and Tanacetum coccineum; rotenone, from the roots of Derris species and related plants; and oils, from petroleum. Though these compounds were originally derived mainly from plant extracts, the toxic agents of some of them (e.g., pyrethrins) have been synthesized. Natural insecticides are usually short-lived on plants and cannot provide protection against prolonged invasions. Except for pyrethrum, they have largely been replaced by newer synthetic organic insecticides.

Fumigants are toxic compounds that enter the respiratory system of the insect through its spiracles, or breathing openings. They include such chemicals as hydrogen cyanide, naphthalene, nicotine, and methyl bromide and are used mainly for killing insect pests of stored products or for fumigating nursery stock.

Synthetic Insecticides

The synthetic contact insecticides are now the primary agents of insect control. In general they penetrate insects readily and are toxic to a wide range of species. The main synthetic groups are the chlorinated hydrocarbons, organic phosphates (organophosphates), and carbamates.

Chlorinated Hydrocarbons

The chlorinated hydrocarbons were developed beginning in the 1940s after the discovery of the insecticidal properties of DDT. Other examples of this series are BHC, lindane, Chlorobenzilate, methoxychlor, and the cyclodienes (which include aldrin, dieldrin, chlordane, heptachlor, and endrin). Some of these compounds are quite stable and have a long residual action; they are, therefore, particularly valuable where protection is required for long periods. Their toxic action is not fully understood, but they are known to disrupt the nervous system. A number of these insecticides have been banned for their deleterious effects on the environment.

Organophosphates

The organophosphates are now the largest and most versatile class of insecticides. Two widely used compounds in this class are parathion and malathion; others are Diazinon, naled, methyl parathion, and dichlorvos. They are especially effective against sucking insects such as aphids and mites, which feed on plant juices. The chemicals' absorption into the plant is achieved either by spraying the leaves or by applying solutions impregnated with the chemicals to the soil, so that intake occurs through the roots. The organophosphates usually have little residual action and are important, therefore, where residual tolerances limit the choice of insecticides. They are generally much more toxic than the chlorinated hydrocarbons. Organophosphates kill insects by inhibiting the enzyme cholinesterase, which is essential in the functioning of the nervous system.

Carbamates

The carbamates are a group of insecticides that includes such compounds as carbamyl, methomyl, and carbofuran. They are rapidly detoxified and eliminated from animal tissues. Their toxicity is thought to arise from a mechanism somewhat similar to that for the organophosphates.

Environmental Contamination and Resistance

The advent of synthetic insecticides in the mid-20th century made the control of insects and other arthropod pests much more effective, and such chemicals remain essential in modern agriculture despite their environmental drawbacks. By preventing crop losses, raising the quality of produce, and lowering the cost of farming, modern insecticides increased crop yields by as much as 50 percent in some regions of the world in the period 1945–65. They have also been important in improving the health of both humans and domestic animals; malaria, yellow fever, and typhus, among other infectious diseases, have been greatly reduced in many areas of the world through their use.

But the use of insecticides has also resulted in several serious problems, chief among them environmental contamination and the development of resistance in pest species. Because insecticides are poisonous compounds, they may adversely affect other organisms besides harmful insects. The accumulation of some insecticides in the environment can in fact pose a serious threat to both wildlife and humans. Many insecticides are short-lived or are metabolized by the animals that ingest them, but some are persistent, and when applied in large amounts they pervade the environment. When an insecticide is applied, much of it reaches the soil, and groundwater can become contaminated from direct application or runoff from treated areas. The main soil contaminants are the chlorinated hydrocarbons such as DDT, aldrin, dieldrin, heptachlor, and BHC. Owing to repeated sprayings, these chemicals can accumulate in soils in surprisingly large amounts (10–112 kilograms per hectare [10–100 pounds per acre]), and their effect on wildlife is greatly increased as they become associated with food chains. The stability of DDT and its relatives leads to their accumulation in the bodily tissues of insects that constitute the diet of other animals higher up the food chain, with toxic effects on the latter. Birds of prey such as eagles, hawks, and falcons are usually most severely affected, and serious declines in their populations have been traced to the effects of DDT and its relatives. Consequently, the use of such chemicals began to be restricted in the 1960s and banned outright in the 1970s in many countries.

Cases of insecticide poisoning of humans also occur occasionally, and the use of one common organophosphate, parathion, was drastically curtailed in the United States in 1991 owing to its toxic effects on farm labourers who were directly exposed to it.

Another problem with insecticides is the tendency of some target insect populations to develop resistance as their susceptible members are killed off and those resistant strains that survive multiply, eventually perhaps to form a majority of the population. Resistance denotes a formerly susceptible insect population that can no longer be controlled by a pesticide at normally recommended rates. Hundreds of species of harmful insects have acquired resistance to different synthetic organic pesticides, and strains that become resistant to one insecticide may also be resistant to a second that has a similar mode of action to the first. Once resistance has developed, it tends to persist in the absence of the pesticide for varying amounts of time, depending on the type of resistance and the species of pest.

Insecticides may also encourage the growth of harmful insect populations by eliminating the natural enemies that previously held them in check. The nonspecific nature of broad-spectrum chemicals makes them more likely to have such unintended effects on the abundance of both harmful and beneficial insects.

Because of the problems associated with the heavy use of some chemical insecticides, current insect-control practice combines their use with biological methods in an approach called integrated control. In this approach, a minimal use of insecticide may be combined with the use of pest-resistant crop varieties; the use of crop-raising methods that inhibit pest proliferation; the release of organisms that are predators or parasites of the pest species; and the disruption of the pest's reproduction by the release of sterilized pests.

Type of Activity

Systemic insecticides become incorporated and distributed systemically throughout the whole plant. When insects feed on the plant, they ingest the insecticide. Systemic insecticides produced by transgenic plants are called plant-incorporated protectants (PIPs). For instance, a gene that codes for a specific Bacillus thuringiensis biocidal protein was introduced into corn (maize) and other species. The plant manufactures the protein, which kills the insect when consumed.

Contact insecticides are toxic to insects upon direct contact. These can be inorganic insecticides, which are metals and include the commonly used sulfur, and the less commonly used arsenates, copper and fluorine compounds. Contact insecticides can also be organic insecticides, i.e. organic chemical compounds, synthetically produced, and comprising the largest numbers of pesticides used today. Or they can be natural compounds like pyrethrum, neem oil etc. Contact insecticides usually have no residual activity.

Efficacy can be related to the quality of pesticide application, with small droplets, such as aerosols often improving performance.

Synthetic Insecticide and Natural Insecticides

A major emphasis of organic chemistry is the development of chemical tools to enhance agricultural productivity. Insecticides represent a major area of emphasis. Many of the major insecticides are inspired by biological analogues. Many others are completely alien to nature.

Organochlorides

The best known organochloride, DDT, was created by Swiss scientist Paul Müller. For this discovery, he was awarded the 1948 Nobel Prize for Physiology or Medicine. DDT was introduced in 1944. It functions by opening sodium channels in the insect's nerve cells. The contemporaneous rise of the chemical industry facilitated large-scale production of DDT and related chlorinated hydrocarbons.

Organophosphates and Carbamates

Organophosphates are another large class of contact insecticides. These also target the insect's nervous system. Organophosphates interfere with the enzymes acetylcholinesterase and other cholinesterases, disrupting nerve impulses and killing or disabling the insect. Organophosphate insecticides and chemical warfare nerve agents (such as sarin, tabun, soman, and VX) work in the same way. Organophosphates have a cumulative toxic effect to wildlife, so multiple exposures to the chemicals amplifies the toxicity. In the US, organophosphate use declined with the rise of substitutes.

Carbamate insecticides have similar mechanisms to organophosphates, but have a much shorter duration of action and are somewhat less toxic.

Pyrethroids

Pyrethroid pesticides mimic the insecticidal activity of the natural compound pyrethrum, the biopesticide found in pyrethrins. These compounds are nonpersistent sodium channel modulators and are less toxic than organophosphates and carbamates. Compounds in this group are often applied against household pests.

Neonicotinoids

Neonicotinoids are synthetic analogues of the natural insecticide nicotine (with much lower acute mammalian toxicity and greater field persistence). These chemicals are acetylcholine receptor agonists. They are broad-spectrum systemic insecticides, with rapid action (minutes-hours). They are applied as sprays, drenches, seed and soil treatments. Treated insects exhibit leg tremors, rapid wing motion, stylet withdrawal (aphids), disoriented movement, paralysis and death. Imidacloprid may be the most common. It has recently come under scrutiny for allegedly pernicious effects on honeybees and its potential to increase the susceptibility of rice to planthopper attacks.

Ryanoids

Ryanoids are synthetic analogues with the same mode of action as ryanodine, a naturally occurring insecticide extracted from Ryania speciosa (Flacourtiaceae). They bind to calcium channels in cardiac and skeletal muscle, blocking nerve transmission. The first insecticide from this class to be registered was Rynaxypyr, generic name chlorantraniliprole.

TYPES OF INSECTICIDES

Organochlorines

The organochlorines are insecticides that contain carbon (thus organo-), hydrogen, and chlorine. They are also known by other names: chlorinated hydrocarbons, chlorinated organics, chlorinated insecticides, and chlorinated synthetics. The organochlorines are now primarily of historic interest, since few survive in today's arsenal.

Diphenyl Aliphatics

The oldest group of the organochlorines is the diphenyl aliphatics, which included DDT, DDD, dicofol, ethylan, chlorobenzilate, and methoxychlor. DDT is probably the best known and most notorious chemical of the 20th century. It is also fascinating, and remains to be acknowledged as the most useful insecticide developed. More than 4 billion pounds of DDT were used throughout the world, beginning in 1940, and in the U.S. ending essentially in 1973, when the U.S. Environmental Protection Agency canceled all uses. The remaining First World countries rapidly followed suit. DDT is still effectively used for malaria control in several third world countries. In 1948, Dr. Paul Muller, a Swiss entomologist, was awarded the Nobel Prize in Medicine for his lifesaving discovery of DDT as an insecticide useful in the control of malaria, yellow fever and many other insect-vectored diseases.

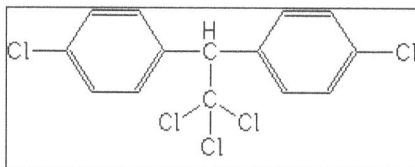

Mode of action: The mode of action for DDT has never been clearly established, but in some complex manner it destroys the delicate balance of sodium and potassium ions within the axons of the neuron in a way that prevents normal transmission of nerve impulses, both in insects and mammals. It apparently acts on the sodium channel to cause "leakage" of sodium ions. Eventually the affected neurons fire impulses spontaneously, causing the muscles to twitch "DDT jitters" followed by convulsions and death. DDT has a negative temperature correlation the lower the surrounding temperature the more toxic it becomes to insects.

Hexchlorocyclohexane (HCH)

Also known as benzenehexachloride (BHC), the insecticidal properties of HCH were discovered in 1940 by French and British entomologists. In its technical grade, there are five isomers, *alpha, beta, gamma, delta* and *epsilon*. Surprisingly, only the *gamma* isomer has insecticidal properties. Consequently, the *gamma* isomer was isolated in manufacture and sold as the odorless insecticide *lindane*. In contrast, technical grade HCH has a strong musty odor and flavor, which can be imparted to treated crops and animal products. Because of its very low cost, HCH is still used in many developing countries. In 2002, the U.S. EPA removed all food-related (tolerance-requiring) uses of lindane from the U.S.

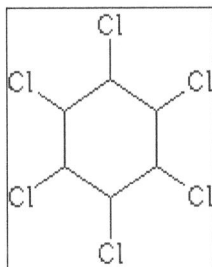

Mode of action: The effects of HCH superficially resemble those of DDT, but occur much more rapidly, and result in a much higher rate of respiration in insects. The *gamma* isomer is a neurotoxicant whose effects arc normally sccn within hours as increased activity, tremors, and convulsions leading to prostration. It too, exhibits a negative temperature correlation, but not as pronounced as that of DDT.

Cyclodienes

The cyclodienes appeared after World War II: chlordane, 1945, aldrin and dieldrin, 1948; heptachlor, 1949; endrin, 1951; mirex, 1954; endosulfan, 1956; and chlordecone, 1958. There were other cyclodienes of minor importance developed in the U.S. and Germany. Most of the cyclodienes are persistent insecticides and are stable in soil and relatively stable to the ultraviolet of sunlight. As a result, they were used in greatest quantity as soil insecticides (especially chlordane, heptachlor, aldrin, and dieldrin) for the control of termites and soil-borne insects whose larval stages feed on the roots of plants.

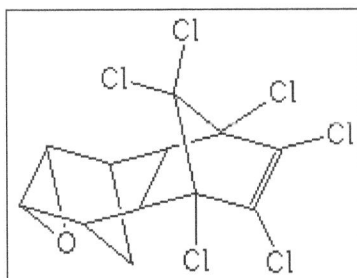

To appreciate the effectiveness of these materials as termiticides, consider that wood and wooden structures treated with chlordane, aldrin, and dieldrin in the year of their development are still protected from damage—after more than 60 years. The cyclodienes were the most effective, long-lasting and economical termiticides ever developed. Because of their persistence in the environment, resistance that developed in several soil insect pests, and in some instances *biomagnification* in wildlife food chains, most agricultural uses of cyclodienes were canceled by the EPA between 1975 and 1980, and their use as termiticides canceled in 1984-88.

Mode of action: Unlike DDT and HCH, the cyclodienes have a positive temperature correlation - their toxicity increases with increasing ambient temperature. Their modes of action are also not clearly understood. However, it is known that this group acts on the inhibitory mechanism called the GABA (g-aminobutyric acid) receptor. This receptor operates by increasing chloride ion permeability of neurons. Cyclodienes prevent chloride ions from entering the neurons, and thereby antagonize the "calming" effects of GABA. Cyclodienes appear to affect all animals similarly, first with the nervous activity followed by tremors, convulsions and prostration.

Polychloroterpenes

Only two polychloroterpenes were developed - toxaphene in 1947, and strobane in 1951. Toxaphene had by far the greatest use of any single insecticide in agriculture, while strobane was relatively insignificant. Toxaphene was used on cotton, first in combination with DDT, for alone it had minimal insecticidal qualities. Then, in 1965, after several major cotton insects became resistant to DDT, toxaphene was formulated with methyl parathion, an organophosphate insecticide.

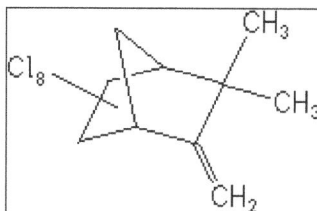

Toxaphene is a mixture of more than 177 10-carbon polychlorinated derivatives. These materials persist in the soil, though not as long as the cyclodienes, and disappear from the surfaces of plants in 3-4 weeks. This disappearance is attributed more to volatility than to photolysis or plant metabolism. Toxaphene is rather easily metabolized by mammals and birds, and is not stored in body fat nearly to the extent of DDT, HCH and the cyclodienes. Despite its low toxicity to insects, mammals and birds, fish are highly susceptible to toxaphene poisoning, in the same order of magnitude as to the cyclodienes. Toxaphene's registrations were canceled by EPA in 1983.

Mode of action: Toxaphene and strobane act on the neurons, causing an imbalance in sodium and potassium ions, similar to that of the cyclodiene insecticides.

Organophosphates

Organophosphates (OPs) is the term that includes all insecticides containing phosphorus. Other names used, but no longer in vogue, are *organic phosphates, phosphorus insecticides, nerve gas relatives,* and *phosphoric acid esters.* All organophosphates are derived from one of the phosphorus acids, and as a class are generally the most toxic of all pesticides to vertebrates. Because of the similarity of OP chemical structures to the "nerve gases," their modes of action are also similar. Their insecticidal qualities were first observed in Germany during World War II in the study of the extremely toxic OP nerve gases *sarin, soman,* and *tabun.* Initially, the discovery was made in search of substitutes for nicotine, which was heavily used as an insecticide but in short supply in Germany.

The OPs have two distinctive features: they are generally more toxic to vertebrates than other classes of insecticides, and most are chemically unstable or nonpersistent. It is this latter characteristic that brought them into agricultural use as substitutes for the persistent *organochorines.* Because of the relatively high toxicity of the OP's, EPA, under provisions of the Food Quality Protection Act, undertook an extensive reappraisal of the entire class beginning in the late 1990's. Many OP's were voluntarily canceled and others lost uses.

Mode of action: The OPs work by inhibiting certain important enzymes of the nervous system, namely cholinesterase (ChE). The enzyme is said to be phosphorylated when it becomes attached to the phosphorous moiety of the insecticide, a binding that is irreversible. This inhibition results in the accumulation of acetylcholine (ACh) at the neuron/neuron and neuron/muscle (neuromuscular) junctions or synapses, causing rapid twitching of voluntary muscles and finally paralysis.

Classification

All OPs are estersof phosphorus having varying combinations of oxygen, carbon, sulfur and nitrogen attached, resulting in six different subclasses: phosphates, phospho-nates, phosphorothioates, phosphorodithioates, phosphorothiolates and phosphoramidates. These subclasses are easily identified by their chemical names.

The OPs are generally divided into three groups- *aliphatic, phenyl,* and *heterocyclic* derivatives.

Aliphatics

The aliphatic OPs are carbon chain-like in structure. The first OP brought to agriculture, TEPP belonged to this group. Other examples are malathion, trichlorfon (Dylox), monocrotophos (Azodrin), dimethoate (Cygon), oxydemetonmethyl (Meta Systox), dicrotophos (Bidrin), disulfoton (Di-Syston), dichlorvos (Vapona), mevinphos (Phosdrin), methamidophos (Monitor), and acephate (Orthene).

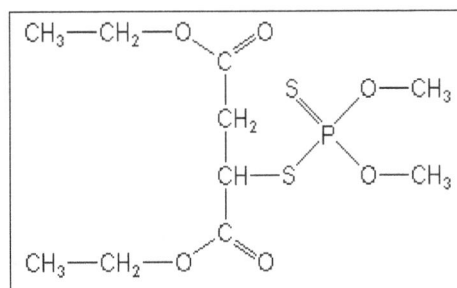

Phenyl Derivatives

The phenyl OPs contain a phenyl ring with one of the ring hydrogens displaced by attachment to the phosphorus moiety and other hydrogens frequently displaced by Cl, NO_2, CH_3, CN, or S. The phenyl OPs are generally more stable than the aliphatics, thus their residues are longer lasting. The first phenyl OP brought into agriculture was parathion (ethyl parathion) in 1947. Examples of other phenyl OPs are methyl parathion, profenofos (Curacron), sulprofos (Bolstar), isofenphos (Oftanol, Pryfon), fenitrothion (Sumithion), fenthion (Dasanit), and famphur (Cyflee. Warbex).

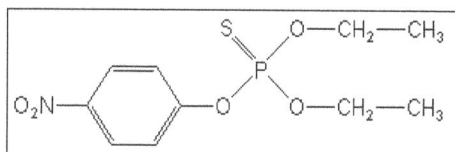

Heterocyclic Derivatives

The term *heterocyclic* means that the ring structures are composed of different or unlike atoms, e.g.,oxygen, nitrogen or sulfur. The first of this group was diazinon introduced in 1952. Other examples in this group are azinphos-methyl (Guthion), azinphos-ethyl (Acifon, Gusathion), chlorpyrifos (Dursban, Lorsban, Lock-On), methidathion (Supracide), phosmet (Imidan), isazophos (Brace, Triumph), and chlorpyrifos-methyl (Reldan).

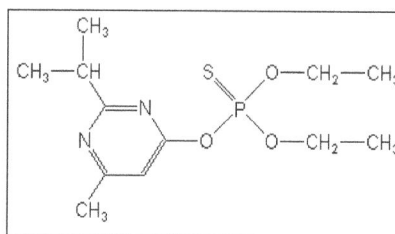

Organosulfurs

These few materials have very low toxicity to insects and are used only as acaricides (miticides). They contain two phenyl rings, resembling DDT, with sulfur in place of carbon as the central atom. These include tetradifon (Tedion), propargite (Omite, Comite), and ovex (Ovotran).

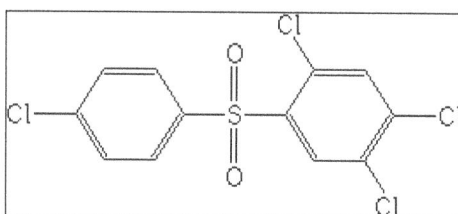

Carbamates

The carbamate insecticides are derivatives of carbamic acid (as the OPs are derivatives of phosphoric acid). And like the OPs, their mode of action is that of inhibiting the vital enzyme *cholinesterase* (ChE).

The first successful carbamate insecticide, carbaryl (Sevin), was introduced in 1956. More of it has been used worldwide than all the remaining carbamates combined. Two distinct qualities have made it the most popular carbamate: its very low mammalian oral and dermal toxicity and an exceptionally broad spectrum of insect control. Other long-standing carbamate insecticides are methomyl (Lannate), carbofuran (Furadan), aldicarb (Temik), oxamyl (Vydate), thiodicarb (Larvin), methiocarb (Mesurol), propoxur (Baygon), bendiocarb (Ficam), carbosulfan (Advantage), aldoxycarb (Standak), promecarb (Carbamult), and fenoxycarb (Logic, Torus). Carbamates more recently introduced include primicarb, indoxacarb, alanycarb and furathiocarb.

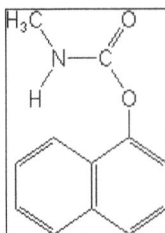

Mode of action: Carbamates inhibit cholinesterase (ChE) as OPs do, and they behave in almost identical manner in biological systems, but with two main differences. Some carbamates are potent inhibitors of aliesterase (miscellaneous aliphatic esterases whose exact functions are not known), and their selectivity is sometimes more pronounced against the ChE of different species. Second, ChE inhibition by carbamates is reversible. When ChE is inhibited by a carbamate, it is said to be *carbamylated*, as when an OP results in the enzyme being *phosphorylated*. In insects, the effects of OPs and carbamates are primarily those of poisoning of the central nervous system, since the insect neuromuscular junction is not cholinergic, as in mammals. The only cholinergic synapses known in insects are in the central nervous system. (The chemical neuromuscular junction transmitter in insects is thought to be glutamic acid.)

Formamidines

The formamidines comprise a small group of insecticides. Three examples are chlordimeform (Galecron, Fundal), which is no longer registered in the U.S., formetanate (Carzol), and amitraz (Mitac, Ovasyn). Their current value lies in the control of OP- and carbamate-resistant pests.

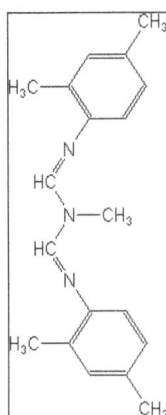

Mode of action: Formamidine poisoning symptoms are distinctly different from other insecticides. Their proposed action is the inhibition of the enzyme monoamine oxidase, which is responsible for degrading the neurotransmitters norepinephrine and serotonin. This results in the accumulation of these compounds, which are known as biogenic amines. Affected insects become quiescent and die.

Dinitrophenols

The basic dinitrophenol molecule has a broad range of toxicities - as herbicides, insecticides, ovicides, and fungicides. Of the insecticides, binapacryl (Morocide) and dinocap (Karathane) were the most recently used. Dinocap is an effective miticide and was very heavily used as a fungicide for the control of powdery mildew fungi. Because of the inherent toxicity of the dinitrophenols, they have all been withdrawn.

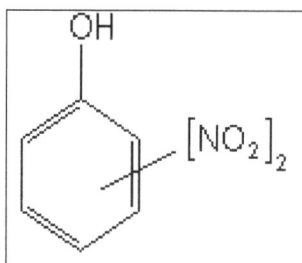

Mode of action: Dinitrophenols act by uncoupling or inhibiting oxidative phosphorylation, which basically prevents the formation of the high-energy phosphate molecule, adenosine triphosphate (ATP).

Organotins

The organotins are a group of acaricides that double as fungicides. Of particular interest is cyhexatin (Plictran), one of the most selective acaricides known, introduced in 1967. Fenbutatin-oxide (Vendex) has been used extensively against mites on deciduous fruits, citrus, greenhouse crops, and ornamentals.

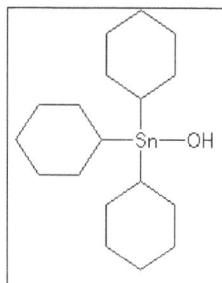

Mode of action: These ten compounds inhibit oxidative phosphorylation at the site of dinitrophenol uncoupling, preventing the formation of the high-energy phosphate molecule adenosine triphosphate (ATP). These trialkyl tins also inhibit photophosphorylation in chloroplasts, the chlorophyll-bearing subcellular units) and could therefore serve as algicides.

Pyrethroids

Natural pyrethrum has seldom been used for agricultural purposes because of its cost and instability in sunlight. In recent decades, many synthetic pyrethrin-like materials have become available. They were originally referred to as *synthetic pyrethroids*. Currently, the better nomenclature is simply *pyrethroids*. These are stable in sunlight and are generally effective against most agricultural insect pests when used at the very low rates of 0.01 to 0.1 pound per acre.

The pyrethroids have an interesting evolution, which is conveniently divided into four generations. The first generation contains only one pyrethroid, allethrin (Pynamin), which appeared in 1949. Its synthesis was very complex, involving 22 chemical reactions to reach the final product.

The second generation includes tetramethrin (Neo-Pynamin), followed by resmethrin (Synthrin) in 1967 (20X as effective as pyrethrum), then bioresmethrin (50X as effective as pyrethrum), then Bioallethrin, and finally phonothrin (Sumithrin).

The third generation includes fenvalerate and permethrin which appeared in 1972-73. These became the first agricultural pyrethroids because of their exceptional insecticidal activity (0.1 lb ai/A) and their photostability. They were virtually unaffected by ultraviolet in sunlight, lasting 4-7 days as efficacious residues on crop foliage.

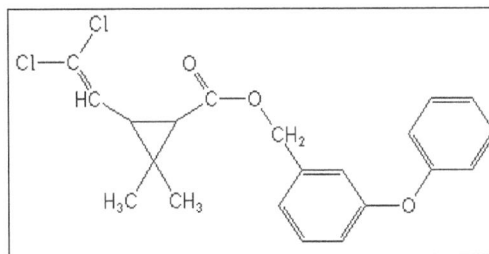

The fourth and current generation, is truly exciting because of their effectiveness in the range of 0.01 to 0.05 lb ai/A. These include bifenthrin (Capture, Talstar), *lambda*-cyhalothrin, cypermethrin, cyfluthrin, deltamethrin (Decis) esfenvalerate (Asana, Hallmark), fenpropathrin (Danitol), flucythrinate (Cybolt, Payoff), fluvalinate (Mavrik, Spur, discontinued), prallethrin (Etoc), *tau*-fluvalinate (Mavrik) tefluthrin, tralomethrin (Scout X-TRA, Tralex), and *zeta*-cypermethrin. All of these are photostable, that is, they do not undergo photolysis (splitting) in sunlight. And because they have minimal volatility they provide extended residual effectiveness, up to 10 days under optimum conditions.

Recent additions to the fourthgeneration pyrethroids are acrinathrin (Rufast), imiprothrin (Pralle), registered in 1998, and *gamma*-cyhalothrim (Pytech), which is in development.

Mode of action: The pyrethroids share similar modes of action, resembling that of DDT, and are considered axonic poisons. They apparently work by keeping open the sodium channels in neuronal membranes. There are two types of pyrethroids. Type I, among other physiological responses, have a negative temperature coefficient, resembling that of DDT. Type II, in contrast have a positive temperature coefficient, showing increased kill with increase in ambient temperature. Pyrethroids affect both the peripheral and central nervous system of the insect. They initially stimulate nerve cells to produce repetitive discharges and eventually cause paralysis. Such effects are caused by their action on the sodium channel, a tiny hole through which sodium ions are permitted to enter the axon to cause excitation. The stimulating effect of pyrethroids is much more pronounced than that of DDT.

Nicotinoids

The nicotinoids are a newer class of insecticides with a new mode of action. They have been previously referred to as *nitro-quanidines, neonicotinyls, neonicotinoids, chloronicotines,* and more recently as the *chloronicotinyls*. Just as the synthetic pyrethroids are similar to and modeled after the natural pyrethrins, so too, are the nicotinoids similar to and modeled after the natural nicotine. Imidacloprid was introduced in Europe and Japan in 1990 and first registered in the U.S. in 1992. It is currently marketed as several proprietary products worldwide, e.g., Admire, Confidor, Gaucho, Merit, Premier, Premise and Provado. Very possibly it is used in the greatest volume globally of all insecticides.

Imidacloprid is a systemic insecticide, having good root-systemic characteristics and notable contact and stomach action. It is used as a soil, seed or foliar treatment in cotton, rice cereals, peanuts, potatoes, vegetables, pome fruits, pecans and turf, for the control of sucking insects, soil insects, whiteflies, termites, turf insects and the Colorado potato beetle, with long residual control. Imidacloprid has no effect on mites or nematodes.

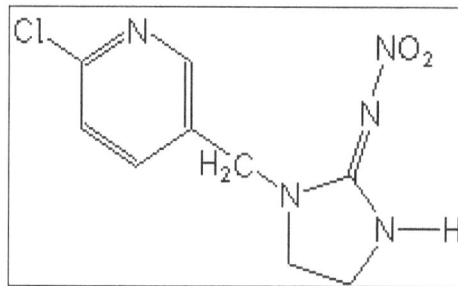

Other nicotinoids include acetamiprid (Assail), thiamethoxam (Actara, Platinum), nitenpyram (Bestguard), clothianidin (Poncho, dinotefuran (Starke) and thiacloprid).

Mode of action: The nicotinoids act on the central nervous system of insects, causing irreversible blockage of postsynaptic nicotinergic acetylcholine receptors.

Spinosyns

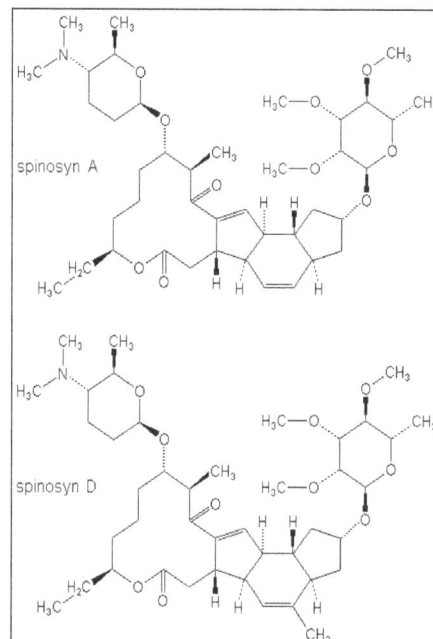

Spinosyns are among the newest classes of insecticides, represented by spinosad. Spinosad is a fermentation metabolite of the actinomycete *Saccharopolyspora spinosa*, a soil-inhabiting microorganism. It has a novel molecular structure and mode of action that provide excellent crop protection typically associated with synthetic insecticides, first registered for use on cotton in 1997. Spinosad is a mixture of spinosyns A and D (thus its name, spinosAD). It is particularly effective as a broad-spectrum material for most caterpillar pests at the astonishing rates of 0.04 to 0.09 pound of active ingredient (18 to 40 grams) per acre. It has both contact and stomach activity against lepidopteran larvae, leaf miners, thrips, and termites, with long residual activity. Crops registered include cotton, vegetables, tree fruits, ornamentals and others.

Mode of action: Spinosad acts by disrupting binding of acetylcholine in nicotinic acetylcholine receptors at the postsynaptic cell.

Fiproles or Phenylpyrazoles

Fipronil (Regent, Icon, Frontline) is the only insecticide in this new class, introduced in 1990 and registered in the U.S. in 1996. It is a systemic material with contact and stomach activity. Fipronil is used for the control of many soil and foliar insects, (e.g., corn rootworm, Colorado potato beetle, and rice water weevil) on a variety of crops, primarily corn, turf, and for public health insect control. It is also used for seed treatment and formulated as baits for cockroaches, ants and termites. Fipronil is effective against insects resistant or tolerant to pyrethroid, organophosphate and carbamate insecticides.

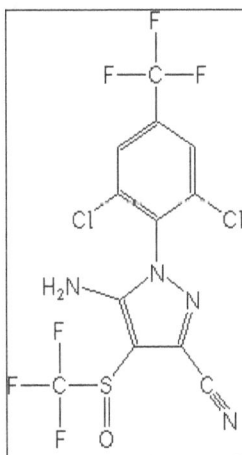

Mode of action: Fipronil blocks the (g-aminobutyric acid- (GABA)) regulated chloride channel in neurons, thus antagonizing the "calming" effects of GABA, similar to the action of the Cyclodienes.

Pyrroles

Chlorfenapyr (Alert, Pirate) is the first and only member of this unique chemical group, as both a contact and stomach insecticide-miticide. It is used on cotton and experimentally on corn, soybeans, vegetables, tree and vine crops, and ornamentals to control whitefly, thrips, caterpillars, mites, leafminers, aphids, and Colorado potato beetle. It has ovicidal activity on some species. EPA took the unusual step of refusing to register chlorfenapyr in 2000 for cotton insect control because of potential hazards to birds. However, labels for greenhouse ornamentals were granted in 2001.

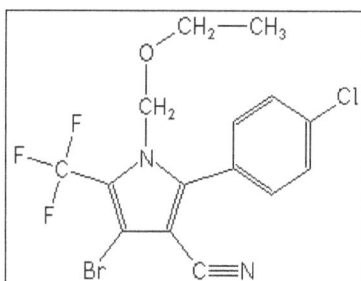

Mode of action: Chlorfenapyr is an "uncoupler" or inhibitor of oxidative phosphorylation, preventing the formation of the crucial energy molecule adenosine triphosphate (ATP).

Pyrazoles

The original pyrazoles were tebufenpyrad and fenpyroximate. These were designed primarily as non-systemic contact and stomach miticides, but do have limited effectiveness on psylla, aphids, whitefly, and thrips. Tebufenpyrad (Pyranica, Masai), registered by EPA in 2002, is used on cotton, soybeans, vegetables, pome fruits, grapes and citrus. Fenpyroximate (Acaban, Dynamite) controls all stages of mites, gives fast knockdown, inhibits molting of immature stages of mites, and has long residual activity. Newer members of this class include ethiprole (Curbex) which is active on a broad sprectum of chewing and sucking insects, and tolfenpyrad (OMI-88) which is reputed to active on pests infesting cole and cucurbit crops.

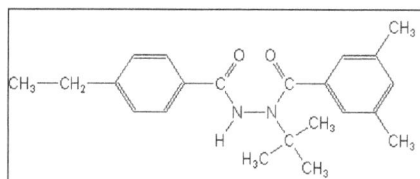

Mode of action: Their mode of action is that of inhibiting mitochondrial electron transport at the NADH-CoQ reductase site, leading to the disruption of adenosine triphosphate (ATP) formation, the crucial energy molecule.

Pyridazinones

Pyridaben (Nexter, Sanmite) is the only member of this class. It is a selective contact insecticide and miticide, also effective against thrips, aphids, whiteflies and leafhoppprs. Registrations are for pome fruits, almonds, citrus, ornamentals and greenhouse ornamentals. Pyridaben provides exceptionally long residual control, and rapid knockdown at a broad range of temperatures.

Mode of action: Pyridaben is a metabolic inhibitor that interrupts mitochondrial electron transport at Site 1, similar to the Quinazolines, below.

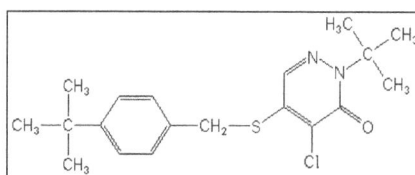

Quinazolines

The quinazolines offer a unique chemical configuration, consisting only of one insecticide, fenazaquin (Matador). Fenazaquin is a contact and stomach miticide. It has ovicidal activity, gives rapid knockdown, and controls all stages of mites. Not yet registered in the U.S., it is used on cotton, stone and pome fruits, citrus, grapes and ornamentals.

Fenazaquin (Matador)

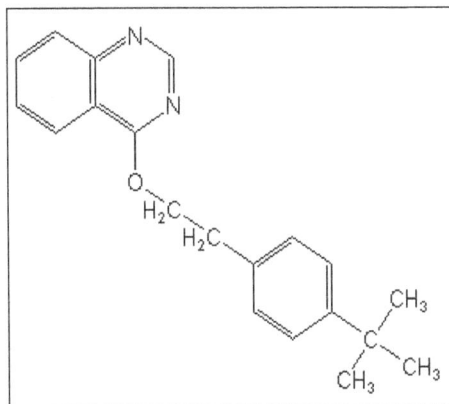

Mode of action: Fenazaquin inhibits mitochondrial electron transport at Site 1, similar to the Pyridazinones, above.

Benzoylureas

Benzoylureas are an entirely different class of insecticides that act as insect growth regulators (IGRs). Rather than being the typical poisons that attack the insect nervous system, they interfere with chitin synthesis and are taken up more by ingestion than by contact. Their greatest value is in the control of caterpillars and beetle larvae.

Benzoylureas were first used in Central America in 1985, to control a severe, resistant leafworm complex (*Spodoptera spp., Trichoplusia spp.*) outbreak in cotton. The withdrawal of the ovicide chlordimeform made their control quite difficult due to their high resistance to almost all insecticide classes, including the pyrethroids.

The benzoylureas were introduced in 1978 by Bayer of Germany, triflumuron (Alsystin) being the first. Others appearing since then are chlorfluazuron (Atabron), followed by teflubenzuron (Nomolt, Dart), hexaflumuron (Trueno, Consult), flufenoxuron (Cascade), and flucycloxuron (Andalin). Others are flurazuron, novaluron, and diafenthiuron, bistrifluron (DBI-3204) and noviflumuron (XDE-007). Until recently lufenuron (Axor) was the newest addition to this group, appearing in 1990. Among the newer benzoylureas only hexaflumuron and novaluron have been registered by EPA.

The only other benzoylurea registered in the U.S. is diflubenzuron (Dimilin, Adept, Micromite). It was first registered in 1982 for gypsy moth, cotton boll weevil, most forest caterpillars, soybean caterpillars, and mushroom flies, but now with a much broader range of registrations.

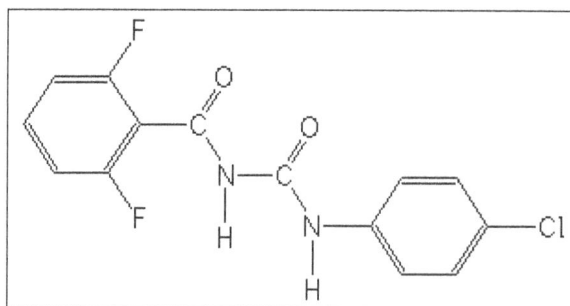

Though not a benzoylurea, cyromazine (Larvadex, Trigard), a triazine, is also a potent chitin synthesis inhibitor. It is selective toward Dipterous species and used for the control of leafminers in vegetable crops and ornamentals, and fed to poultry or sprayed to control flies in manure of broiler and egg producing operations, and incorporated into compost of mushroom houses for fungus gnats.

Mode of action: The benzoylureas act on the larval stages of most insects by inhibiting or blocking the synthesis of chitin, a vital and almost indestructible part of the insect exoskeleton. Typical effects on developing larvae are the rupture of malformed cuticle or death by starvation. Adult female boll weevils exposed to diflubenzuron lay eggs that do not hatch. And, mosquito larvae control can be achieved with as little as 1.0 gram of diflubenzuron per acre of surface water.

Botanicals

Botanical insecticides are of great interest to many, for they are *natural* insecticides, toxicants derived from plants. Historically, the plant materials have been in use longer than any other group, with the possible exception of sulfur. Tobacco, pyrethrum, derris, hellebore, quassia, camphor, and turpentine were some of the more important plant products in use before the organized search for insecticides began in the early 1940s.

In recent years the term *biorational* has been put into play by the EPA. There are similarities and differences between the terms botanical and biorational.

Botanical insecticide use in the U.S. peaked in 1966, and has declined steadily since. Pyrethrum is now the only classical botanical of significance in use. Some newer plant-derived insecticides that have come into use are referred to as *florals or scented plant chemicals* and include, among others, limonene, cinnamaldehyde and eugenol. In addition, there is azadirachtin from the neem tree which is used in greenhouse and on ornamentals.

Pyrethrum

Pyrethrum is extracted from the flowers of a chrysanthemum grown in Kenya and Ecuador. It is one of the oldest and safest insecticides available. The ground, dried flowers were used in the early 19th century as the original louse powder to control body lice in the Napoleonic Wars. Pyrethrum acts on insects with phenomenal speed causing immediate paralysis, thus its popularity in fast knockdown household aerosols. However, unless it is formulated with one of the *synergists*, most of the paralyzed insects recover to once again become pests. Pyrethrum is a mixture of four compounds: pyrethrins I and II and cinerins I and II.

Mode of action: Pyrethrum is an axonic poison, as are the synthetic pyrethroids and DDT. Axonic poisons are those that in some way affect the electrical impulse transmission along the axons, the elongated extensions of the neuron cell body. Pyrethrum and some pyrethroids have a greater insecticidal effect when the temperature is lowered, a negative temperature coefficient, as does DDT. They affect both the peripheral and central nervous system of the insect. Pyrethrum initially stimulates nerve cells to produce repetitive discharges, leading eventually to paralysis. Such effects are caused by their action on the sodium channel, a tiny hole through which sodium ions are permitted to enter the axon to cause excitation. These effects are produced in insect nerve cord, which contains ganglia and synapses, as well as in giant nerve fiber axons.

Nicotine

Nicotine is extracted by several methods from tobacco, and is effective against most all types of insect pests, but is used particularly for aphids and caterpillars - soft bodied insects. Nicotine is an alkaloid, a chemical class of heterocyclic compounds containing nitrogen and having prominent physiological properties. Other well-known alkaloids that are not insecticides are caffeine (coffee, tea), quinine (cinchona bark), morphine (opium poppy), cocaine (coca leaves), ricinine (a poison in castor oil beans), strychnine (*Strychnos nux vomica*), coniine (spotted hemlock, the poison used by Socrates), and, finally LSD (a hallucinogen from the ergot fungus attacking grain).

Mode of action: Nicotine action is one of the first, classic modes of action identified by pharmacologists. Drugs that act similarly to nicotine are said to have a nicotinic response. Nicotine mimics acetylcholine (ACh) at the neuromuscular (nerve/muscle) junction in mammals, and results in twitching, convulsions, and death, all in rapid order. In insects the same action is observed, but only in the central nervous system ganglia.

Rotenone

Rotenone or rotenoids are produced in the roots of two genera of the legume family: *Derris* and *Lonchocarpus* (also called cubé) grown in South America. It is both a stomach and contact insecticide and used for the last century and a half to control leaf-eating caterpillars, and three centuries prior to that in South America to paralyze fish, causing them to surface and be easily captured. Today, rotenone is used in the same way to reclaim lakes for game fishing. Used on a prescribed basis, it eliminates all fish, closing the lake to reintroduction of rough species. It is a selective piscicide in that it kills all fish at dosages that are relatively nontoxic to fish food organisms, and is degraded rapidly.

Mode of action: Rotenone is a respiratory enzyme inhibitor, acting between NAD+ (a coenzyme involved in oxidation and reduction in metabolic pathways) and coenzyme Q (a respiratory enzyme responsible for carrying electrons in some electron transport chains), resulting in failure of the respiratory functions.

Limonene

Limonene or d-Limonene is the latest addition to the botanicals. Limonene belongs to a group often called *florals* or *scented plant chemicals*. Extracted from citrus peel, it is effective against all external pests of pets, including fleas, lice, mites, and ticks, and is virtually nontoxic to warm-blooded animals. Several insecticidal substances occur in citrus oil, but the most important is limonene, which constitutes about 98% of the orange peel oil by weight. Two other recently introduced floral products are eugenol (Oil of Cloves) and cinnamaldehyde (derived from Ceylon and Chinese cinnamon oils). They are used on ornamentals and many crops to control various insects.

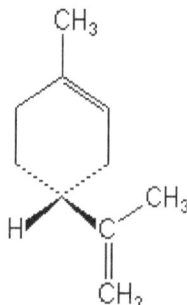

Mode of action: Its mode of action is similar to that of pyrethrum. It affects the sensory nerves of the peripheral nervous system, but it is not a ChE inhibitor.

Neem

Neem oil extracts are squeezed from the seeds of the neem tree and contain the active ingredient *azadirachtin*, a nortriterpenoid belonging to the lemonoids. Azadirachtin has shown some rather sensational insecticidal, fungicidal and bactericidal properties, including insect growth regulating qualities. Azatin is marketed as an insect growth regulator, and Align and Nemix as a stomach/contact insecticide for greenhouse and ornamentals.

Mode of action: Azadirachtin disrupts molting by inhibiting biosynthesis or metabolism of ecdysone, the juvenile molting hormone.

Synergists or Activators

Synergists are not in themselves considered toxic or insecticidal, but are materials used with insecticides to synergize or enhance the activity of the insecticides. The first was introduced in 1940 to increase the effectiveness of pyrethrum. Since then many materials have appeared, but only a few are still marketed. Synergists are found in most all household, livestock and pet aerosols to enhance the action of the fast knockdown insecticides pyrethrum, allethrin, and resmethrin, against flying insects. Current synergists, such as piperonyl butoxide, contain the methylenedioxyphenyl moiety, a molecule found in sesame oil and later named *sesamin*.

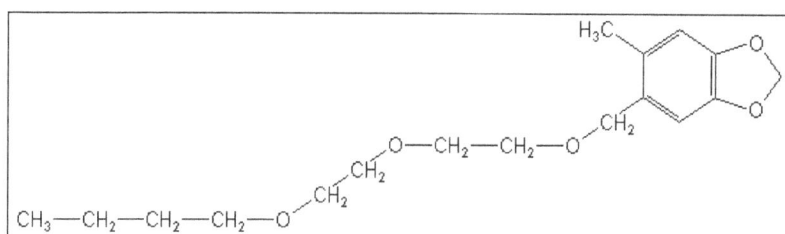

Mode of action: The synergists inhibit cytochrome P-450 dependent polysubstrate monooxygenases (PSMOs), enzymes produced by microsomes, the subcellular units found in the liver of mammals and in some insect tissues (e.g., fat bodies). The earlier name for these enzymes was mixed-function oxidases (MFOs). These PSMOs bind the enzymes that degrade selected foreign substances, such as pyrethrum, allethrin, resmethrin or any other synergized compound. Synergists simply bind the oxidative enzymes and prevent them from degrading the toxicant.

Antibiotics

In this category belong the *avermectins*, which are insecticidal, acaricidal, and antihelminthic agents that have been isolated from the fermentation products of *Streptomyces avermitilis*, a member of the actinomycete family. *Abamectin* is the common name assigned to the avermectins, a mixture of containing 80% avermectin B1*a* and 20% B1*b*, homologs that have about equal biological activity. Clinch is a fire ant bait, and Avid is applied as a miticide/insecticide. Abamectin has certain local systemic qualities, permitting it to kill mites on a leaf's underside when only the upper surface is treated. The most promising uses for these materials are the control of spider mites, leafminers and other difficult-to-control greenhouse pests, and internal parasites of domestic animals.

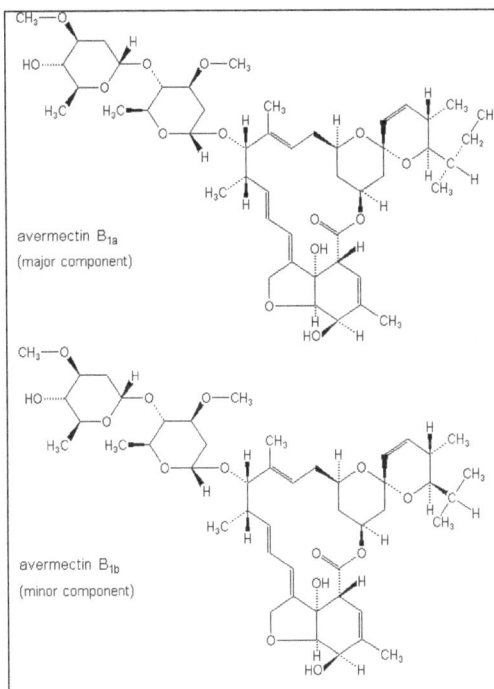

Emamectin benzoate (Proclaim, Denim) is an analog of abamectin, produced by the same fermentation system as abamectin. It was first registered in 1999. It is both a stomach and contact insecticide used primarily for control of caterpillars at the rate of 0.0075 to 0.015 lb (3.5 to 7.0 grams) a.i. per acre. Shortly after exposure, larvae stop feeding and become irreversibly paralyzed, dying in 3-4 days. Rapid photodegradation of both abamectin and emamectin occurs on the leaf surface. More recently, milabectin (Mesa) has been introduced. It is a miticide with activity on piercing/sucking insects and is pending for registration.

Mode of action: Avermectins block the neurotransmitter (g-aminobutyric acid (GABA) at the neuromuscular junction in insects and mites. Visible activity, such as feeding and egg laying, stops shortly after exposure, though death may not occur for several days.

Fumigants

The fumigants are small, volatile, organic molecules that become gases at temperatures above 40 °F. They are usually heavier than air and commonly contain one or more of the halogens (Cl, Br, or F). Most are highly penetrating, reaching through large masses of material. They are used to kill insects, insect eggs, nematodes, and certain microorganisms in buildings, warehouses, grain elevators, soils, and greenhouses and in packaged products such as dried fruits, beans, grain, and breakfast cereals.

Although its use is now in decline because of environmental concerns, methyl bromide is the most heavily used of the fumigants, 68,424 metric tons worldwide in 1996, almost half of which is used in the U.S. The dominant use is for preplanting soil treatments, which accounted for 70% of that global total. Quarantine uses account for 5-8%, while 8% is used to treat perishable products, such as flowers and fruits, and 12% for nonperishable products, like nuts and timber. Approximately 6% is used for structural applications, as in drywood termite fumigation of infested buildings.

With the recently passed change to the Clean Air Act amendments of 1990, U.S. production and importation must be reduced 25% from 1991 levels by 1999. A 50% reduction must be achieved by 2001, followed by a 70% reduction in 2003, and a full ban of the product in 2005. Under the Montreal Protocol, developing countries have until 2015 to phase out methyl bromide production.

Some of the other common fumigants are ethylene dichloride, hydrogen cyanide, sulfuryl fluoride (Vikane), Vapam, TeloneII, D-D, chlorothene, ethylene oxide, and the familiar home-use moth repellents napthalene crystals and paradichlorobenzene crystals.

Phosphine gas (PH_3) has also replaced methyl bromide in a few applications, primarily for insect pests of grain and food commodities. Treatment requires the use of aluminum or magnesium phosphide pellets, which react with atmospheric moisture to produce the gas. Phosphine, however, is very damaging to fresh commodities and is highly adsorbed onto oil, thus does not perform as a soil fumigant.

Alternatives that can fully replace methyl bromide are unlikely to be available by the deadlines set for replacement. Its low cost and utility on a wide variety of pests are hard to match. Because the loss of MeBr has considerable economic consequences, EPA has made it a priority to find and register replacements. To this end some progress has been made. The chemical 1,3-dichloropropane was registered in 2001 for preplant soil fumigation in strawberries and tomatoes. Moreover, iodomethane and metam-potassium are both being evaluated as soil fumigants.

Mode of action: Fumigants, as a group, are narcotics. That is, they act through means more physical than physiological. The fumigants are liposoluble (fat soluble); they have common symptomology; their effects are reversible; and their activity is altered very little by structural changes in their molecules. As narcotics, they induce narcosis, sleep, or unconsciousness, which in effect is their action on insects. Liposolubility appears to be an important factor in their action, since these narcotics lodge in lipid-containing tissues found throughout the insect body, including their nervous system.

Insect Repellents

Historically, repellents have included smoke, plants hung in dwellings or rubbed on the skin as the fresh plant or its brews, oils, pitches, tars, and various earths applied to the body. Before a more edified approach to insect olfaction and behavior was developed, it was wrongly assumed that if a substance was repugnant to humans it would likewise be repellent to annoying insects.

In recent history, the repellents have been dimethyl phthalate, Indalone, Rutgers 612, dibutyl phthalate, various MGK repellents, benzyl benzoate, the military clothing repellent (N-butyl acetanilide), dimethyl carbate (Dimelone) and diethyl toluamide (DEET, Delphene). Of these, only DEET has survived, and is used worldwide for biting flies and mosquitoes. Most of the others have lost their registrations and are no longer available.

In 1999, EPA has registered a new insect repellent, N-methylneodecanamide. Rather than being used on humans to repel insects, it is applied to household floors and other surfaces to repel cockroaches and ants.

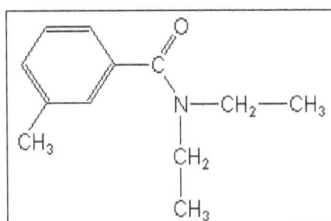

Inorganics

Inorganic insecticides are those that do not contain carbon. Usually they are white crystals in their natural state, resembling the salts. They are stable chemicals, do not evaporate, and are usually water soluble.

Sulfur is very likely the oldest known, effective insecticide. Sulfur and sulfur candles were burned by our great-grandparents for every conceivable purpose, from bedbug fumigation to the cleansing of a house just removed from quarantine of smallpox. Today, sulfur is a highly useful material in integrated pest management programs where target pests specificity is important. Sulfur dusts are especially toxic to mites of every variety, such as chiggers and spider mites, and to thrips and newly-hatched scale insects. Sulfur dusts and sprays are also fungicidal, particularly against powdery mildews.

Several other inorganic compounds have been used as insecticides: mercury, boron, thallium, arsenic, antimony, selenium, and fluoride. Arsenicals have included the copper arsenate, Paris green, lead arsenate, and calcium arsenate. The arsenicals uncouple oxidative phosphorylation, inhibit certain enzymes that contain sulfhydryl (-SH) groups, and coagulate protein by causing the shape or configuration of proteins to change.

The inorganic fluorides were sodium fluoride, barium fluosilicate, sodium silicofluoride, and cryolite (Kryocide). Cryolite has returned in recent years as a relatively safe fruit and vegetable insecticide, used in integrated pest management programs. The fluoride ion inhibits many enzymes that contain iron, calcium, and magnesium. Several of these enzymes are involved in energy production in cells, as in the case of phosphatases and phosphorylases.

Boric acid, used against cockroaches and other crawling household pests in the 1930's and '40's, has also returned. As a salt, it is non-volatile and will remain effective as long as it is kept dry and in adequate concentration. Consequently, it has the longest residual activity of any insecticide used for crawling household insects, and is quite useful in the control of all cockroach species when placed in wall voids and other protected, difficult-to-reach sites. It acts as a stomach poison and insect cuticle wax absorber.

Sodium borate (disodium octaborate tetrahydrate) (Tim-Bor, Bora-Care) resembles boric acid in its action. This water-soluble salt is used to treat lumber and other wood products to control decay fungi, termites, and other wood infesting pests.

The last group of inorganics is the silica gels or silica aerogels - light, white, fluffy, silicate dusts used for household insect control. The silica aerogels kill insects by absorbing waxes from the insect cuticle, permitting the continuous loss of water from the insect body, causing the insects to become desiccated and die from dehydration. These include Dri-Die, Drianone, and Silikil Microcel. Drianone is fortified with pyrethrum and synergists to enhance its effectiveness.

New Miscellaneous Insecticide Classes

Seven classes of insecticides have made their appearance in recent years. These are summarized below.

Methoxyacrylates

Fluacrypyrin (Titaron) is an acaricide for fruit and is the only example of this class currently. It is registered for use on fruit in Japan.

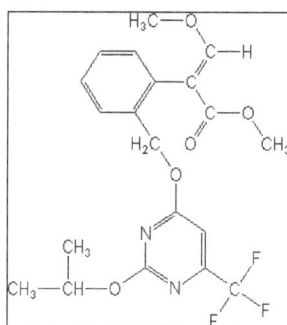

Methyl (*E*)-2-{α-[2-isopropoxy-6-(trifluoromethyl)pyrimidin-4-yloxy]-o-tolyl}-3-methoxyacrylate.

Naphthoquinones

Acequinocyl (Kanemite, Piton) is a miticide with insecticidal activity for pome fruit, nut crops, citrus and ornamentals. It is the only member of this group at present and the mode of action is not yet determined. It holds registrations in Korea and Japan but not in the US.

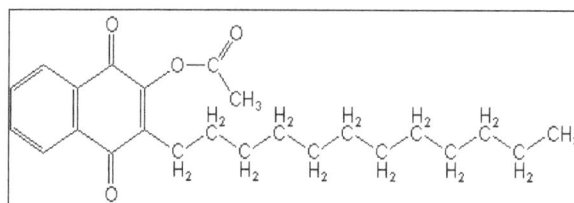

3-dodecyl-1,4-dihydro-1,4-dioxo-2-naphthyl acetate.

Nereistoxin Analogues

These include thiocyclam, cartap, bensultap, and thiocytap-sodium. Analogues of nereistoxin have been known for decades. They generally are stomach poisons with some contact action and often show some systemic action. A major share of the development and use of these compounds has taken place in Japan. They are based on a natural toxin of the marine worm *Lumbriconereis heteropoda*. Of the many analogs synthesized only those that were metabolized back to the original nereistoxin after application were active. In this sense members of this class are *proinsecticides* in that they are applied in their manufactured form but are known to degrade to a specific active component. The members of this group tend to be selectively active on Colopteran and Lepidopteran insect pests. Cartap (Agrotap) is a broad spectrum insecticide with good activity against rice stem borer. Bensultap (Bancol) is used to control the Colorado Potato beetle and other insect pests. Thiosultap-sodium (Pilarhope) is used to control selected beetle and Lepidopteran pests on rice, vegetables and fruit trees.

Thiocyclam (Evisect) is used for the control of similar pests in several crops. Members of this class act as acetyl choline receptor agonists at low concentrations and as channel blockers at higher concentrations. Although there has been commercial interest in thiocyclam for use in the US we do not believe there are commercial examples that are to achieve U.S. registration.

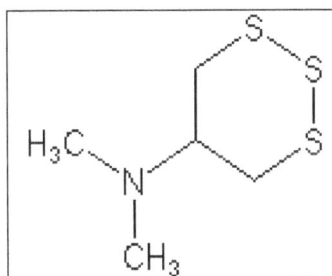

N, N-dimethyl-1,2,3-trithian-5-ylamine.

Pyridine Azomethin

Pymetrozine, first registered in 1999 by EPA has a unique mode of action that is not fully understood. It appears to act by preventing insects from the Order Homoptera from inserting their stylus into plant tissue. Pymetrozine is used to control aphids and whiteflies in vegetables, potatoes, tobacco, deciduous citrus fruit hops and ornamentals.

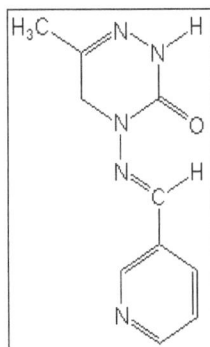

(*E*)-4,5-dihydro-6-methyl-4-(3-pyridylmethyleneamino)-1,2,4-triazin-3(2*H*)-one.

Pyrimidinamines

Pyrimidifen (Miteclean) is an insecticide and miticide. As a miticide the product controls spider and rust mites in deciduous fruits, citrus, vegetables and tea. As an insecticide it controls diamondback moth in vegetables. Very little information is available on the other member of this class, Flufenerim (S-1560), other than it is insecticidal.

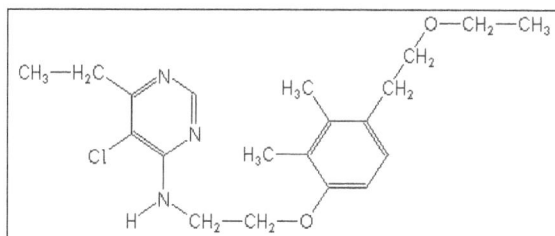

5-chloro-*N*-{2-[4-(2-ethoxyethyl)-2,3-dimethylphenoxy]ethyl}-6-ethylpyrimidin-4-amine.

Tetronic Acids

Spirodiclofen (Envidor) and spiromesifen (Oberon) are the only two members of this recently introduced class. Spirodiclofen has broad-spectrum activity against mites, and controls scale crawlers and psyllad nymphs. Action is good on eggs and quiescent stages. Target crops are citrus, grapes, nuts, pome and stone fruits.

Miscellaneous Compounds

Clofentezine (Apollo, Acaristop), belongs to the unique group, the tetrazines, used as an acaricide/ovicide for deciduous fruits, citrus, cotton, cucurbits, vines and ornamentals. A newer and somewhat similar agent is etoxazole (TerraSan) which is an acaricide registered in 2002 for ornamentals grown in greenhouses. The modes of action of these two compounds are not yet understood.

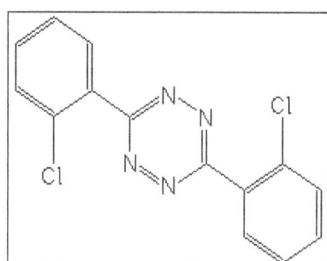

Etoxazole (Terrasan).

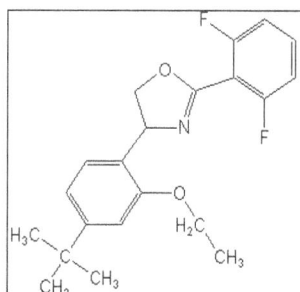

(*RS*)-5-tert-butyl-2-[2-(2,6-difluorophenyl)-4,5-dihydro-1,3-oxazol-4-yl]phenetole.

Enzone, sodium tetrathiocarbonate, is used only on grapes and citrus applied as a water application and irrigated into the soil. It breaks down in the soil to form carbon disulfide, which acts rapidly, decomposes quickly, and is effective against nematodes, soil insects, and soil borne diseases.

The newest agents in this category are pyridanyl and amidoflumet. Pryidalyl (S-1812) is active on Lepidoptera and thrips and has the advantage of being active against pyrethroid-resistant insects. Little more is available on amidoflumet (S-1955) other than it is an acaricide in the early stage of its development.

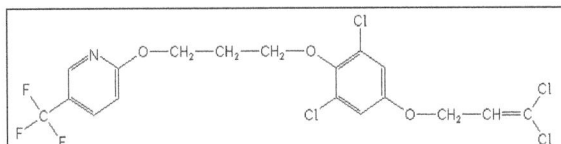

Pyridalyl (S-1812): 2,6-dichloro-4-(3,3-dichloroallyloxy)phenyl 3-[5-(trifluoromethyl)-2-pyridyloxy]propyl ether.

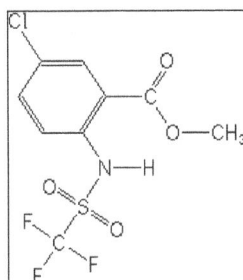

Amidoflumet (S-1955): Methyl 5-chloro-2-{[(trifluoromethyl)sulfonyl]amino}benzoate.

Biorational Insecticides

The U.S. EPA identifies biorational pesticides as inherently different from conventional pesticides, having fundamentally different modes of action, and consequently, lower risks of adverse effects from their use. Biorational has come to mean any substance of natural origin (or man-made substances resembling those of natural origin), that has a detrimental or lethal effect on specific target pest(s), e.g., insects, weeds, plant diseases (including nematodes), and vertebrate pests, possess a unique mode of action, are non-toxic to man and his domestic plants and animals, and have little or no adverse effects on wildlife and the environment. EPA uses a similar term, biopesticides, which will be defined below.

Biorational insecticides are grouped as either (1) biochemicals (hormones, enzymes, pheromones and natural agents, such as insect and plant growth regulators), or (2) microbial (viruses, bacteria, fungi, protozoa, and nematodes). In the 1990's the US-EPA began to emphasize a class of products known as biopesticides. EPA places biopesticides into three categories:

1. Microbial pesticides (bacteria, fungi, virus or protozoa).

2. Biochemicals – natural substances that control pests by non-toxic mechanisms. An example is insect pheromones.

3. Plant-Incorporated protectants (PIPs) – (primarily transgenic plants, e.g., Bt corn).

EPA discloses that at the end of 2001 there were nearly 200 biopesticide active ingredients registered comprising nearly 800 products.

Characteristics that distinguish biorational and biopesticides from conventional ones include: very low orders of toxicity to non-target species, pest targets are specific, generally low use rates, rapid decomposition in the environment, usually work well in IPM programs and reduce reliance on conventional pesticide products.

The terms "biorational" and "biopesticide" overlap but are not identical. In some cases there are overlaps with botanicals (e.g., rotenone, florals, etc. and also conventional insecticides (e.g., benzoylureas). We will point out the discrepancies in classification between the biorational and biopesticide categories where they occur.

Insect Pheromones

Most insects appear to communicate by releasing molecular quantities of highly specific compounds that vaporize readily and are detected by insects of the same species. These delicate molecules are known as pheromones. The word pheromone comes from the Greek pherein, "to carry," and hormon, "to excite or stimulate."

Of 1,314 species of insects with confirmed attraction responses to identified pheromones, 1,260 of these pheromones are produced by females. Only 54 species use male-produced sex attractants. In a few species both sexes produce the same attractant by both sexes.

Pheromones are classified as either, releasers and or primers. Releasers are fast-acting and are used by insects for sexual attraction, aggregation (including trail following), dispersion, oviposition, and alarm. Primers are slow-acting and cause gradual changes in growth and development, especially in social insects by regulating caste ratios of the colony.

The five principal uses for sex pheromones are: (1) male trapping, to reduce the reproductive potential of an insect population; (2) movement studies, to determine how far and where insects move from a given point; (3) population monitoring, to determine when peak emergence or appearance occurs; (4) detection programs, to determine if a pest occurs in a limited trapping area, such as around international airports or quarantined areas; and (5) the "confusion or mating disruption" technique.

The first use of mating disruption involved gossyplure, the pink bollworm pheromone. Incorporated into small, hollow, polyvinyl fibers that permit slow release of the pheromone, it was broadcast heavily and uniformly over infested cotton fields. In mid 2002, EPA had registered 36 pheromones which comprised over 200 individual products.

Despite praise for the potential of sex pheromones, they are most practically used in survey traps to provide information about population levels, to delineate infestations, to monitor control or eradication programs, and to warn of new pest introductions.

Hydrazine Insecticide

A newer class of insecticidal IGRs is the hydrazines, which includes tebufenozide, halofenozide, methoxyfenozide and chromafenozide. All are ecdysone agonists or disruptors. EPA has not classified members of this group as biopesticides. Tebufenozide (Mimic, Confirm), in addition to being both a stomach and contact insecticide, has also JH-IGR characteristics. It disrupts the molting process by antagonizing ecdysone, the molting hormone. Lepidopteran pests are controlled while maintaining natural populations of beneficial insect predators and parasites. Halofenozide registered in 1999, is a systemic IGR, effective on cutworms, sod webworms, armyworms and white grubs, and has some ovicidal activity. It lacks the stomach or contact characteristics of tebufenozide. Methoxyfenozide (Intrepid), like tebufenozide, is both a stomach and contact insecticide with JH-IGR qualities. It is systemic only through the roots. Pests controlled are lepidopterans such as codling moth, oriental fruit moth, European corn borer, and others. Crop candidates are cotton, corn, vegetables, pome fruit, and grapes. EPA considered methoxyfenozide as a reduced-risk candidate and first registered it in mid-2000. Chromafenozide (Matric) is a newer member of this group, not registered in the U.S., and is used to control various lepidopteran pests in vegetables and ornamentals.

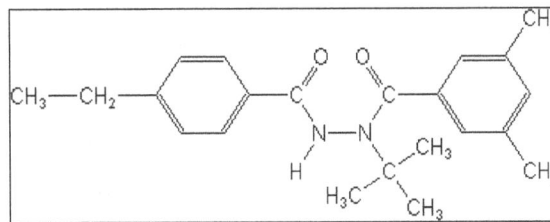

Tebufenozide: 3,5-dimethylbenzoic acid
1-(1,1-dimethylethyl)-2-(4-ethylbenzoyl)hydrazide.

Other Biorational Insecticides

A number of the products that we covered under botanicals and florals are also considered by many to be biorational products, and indeed, EPA includes them under the biopesticide category. Some examples include Neem oil, cinnamaldehyde, and eugenol. A new product, Virtuoso, is a Streptomycetes-based agent that controls caterpillars but little is yet published on it, at present. Clandosan is a naturally occurring product derived from crab and shrimp shells and used as a nematicide. It is a dried, powdered, chitin protein isolated from crustacean exoskeletons and blended with urea. It stimulates growth of beneficial soil microorganisms that control nematodes, but does not have a direct adverse effect on nematodes as such.

Microbials

Microbial insecticides obtain their name from microorganisms that are used to control certain insects. The insect disease-causing microorganisms do not harm other animals or plants. At present there are relatively few produced commercially and approved by the EPA (over 55 natural, and 16 bioengineered organisms) for use on food and feed crops. In mid-2002, the EPA list of registered microbials included 35 bacteria, 1 yeast, 17 fungi, 1 protozoan, 6 viruses, 8 bioengineered organisms and 8 transgenic crop genes.

The insecticidal bacterium *Bacillus thuringiensis* (*Bt*) was discovered in the early 20th century. It occurs as a large number of subspecies that are identified among other characteristics by surface antigens, plasmid arrays, and breadth of species responding to its insecticidal action. *Bt* is a soil inhabiting, gram-positive sporulating bacterium that produces one or more very tiny parasporal crystals within its sporulating cells. These crystals are composed of large proteins known as delta-endotoxins. Delta-endotoxins act by binding to specific receptor sites on the gut epithelium, leading slowly to degradation of the gut lining and starvation. Thus, several days are required to kill insects that have ingested *Bt* products. Over time, several *B. thuringiensis* varieties have been discovered, each with its distinct toxicity characteristics to different insect species. *B. thuringiensis* var. *kurstaki* was the first, being the spores and crystalline delta-endotoxin as the active ingredient, and produced by *B. thuringiensis* Berliner, var. *kurstaki*, Serotype H-3a3b, HD-1, in fermentation. Products from this process control most lepidopteran pests, the caterpillars with high gut pH, which include the armyworms, cabbage looper, imported cabbage worm, gypsy moth, and spruce budworm. The next was *B. thuringiensis* var. *israelensis*, being the crystalline delta-endotoxin as the active ingredient, and produced by fermentation of *B. thuringiensis* Berliner, var. *israelensis*,Serotype H-14. These products are used primarily for the control of aquatic insects, the mosquitoes and black flies in their larval forms. Then came *B. thuringiensis var. aizawai*, produced by this variety, Serotype H-7, in fermentation. This product is currently registered only for the control of the wax moth larval infestations in the honey comb of honey bees. Following this came *B. thuringiensis* var. *morrisoni*, spores and delta-endotoxin produced by fermentation of Serotype 8a8b. This is again a broad spectrum Bt for most caterpillars on most crops including the home garden. *B. thuringiensis* var. *san diego* was developed for Colorado potato beetle control on all its hosts, the elm leaf beetle and other beetle larvae on a wide range of shade and ornamental trees. This was the first Bt product that was effective against coleopteran larvae. *B. thuringiensis* var. *tenebrionis* and the identical var. *san diego*were also developed for the Colorado potato beetle.The utilization of Bt genes transplanted into crops, which is addressed elsewhere in this document, is transforming the area of microbial pesticides.

An innovative development in the agricultural use of microbial insecticides was the addition of feeding or gustatory stimulants, making the mixtures serves as baits. The feeding stimulants attracted the caterpillars to treated foliage, which increased their consumption of the microbial. Two successfully marketed products were Coax and Gustol, both of which were cottonseed meal derivatives formulated as wettable powders. Both products have been discontinued.

- Fungi: Mycar was a promising biorational miticide, a mycoacaricide, but was discontinued by the manufacturer in 1984. The microorganism was *Hirsutella thompsonii*, a parasitic fungus that infects and kills the citrus rust mite. Under optimum conditions *H. thompsonii* can infect spider mites and other nontarget mites. It was, however, consistently effective only against the citrus rust mite; thus a selective miticide.

- EPA registered *Metarhizium anisopliae* St. F52 in mid-2002 to control various ticks, beetles, flies, gnats and thrips for non-food outdoor and greenhouse uses. Certain of the registered uses were conditional for two years pending results of tick performance studies. Another strain of this organism is also registered as a termiticide. Application was made in 1998 to register the fungus. *Aspergillus flavus* strain AF36 as a bioinsecticide for cotton. Its purpose is to help reduce the incidence of other *Aspergillus* spp. that produce the highly toxic mycotoxin, aflatoxin, in cotton seed.

- Protozoa: *Nosema locustae* is a biorational originally developed by Sandoz, Inc., in 1981, for the control of grasshoppers. Marketed under the names NOLO-Bait, NOLO-BB, and Grasshopper Attack, the microorganism is a protozoan. These have been discontinued although the registrations remain.

- Nematodes: There were two commercial nematode products available for termite control, Spear and Saf T-Shield. The nematode, *Neoaplectana carpocapsae*, in the family Steirnernamatidae, is specific for subterranean termites. It kills all stages of these termites by delivering a pathogenic bacterium, *Xenorhabdus* spp., which is lethal within 48 hours after penetration. Unfortunately, neither product succeeded commercially.

DICHLORODIPHENYLTRICHLOROETHANE

Dichlorodiphenyltrichloroethane, commonly known as DDT, is a colorless, tasteless, and almost odorless crystalline chemical compound, an organochlorine, originally developed as an insecticide, and ultimately becoming infamous for its environmental impacts. It was first synthesized in 1874 by the Austrian chemist Othmar Zeidler. DDT's insecticidal action was discovered by the Swiss chemist Paul Hermann Müller in 1939. DDT was used in the second half of World War II to control malaria and typhus among civilians and troops. Müller was awarded the Nobel Prize in Physiology or Medicine "for his discovery of the high efficiency of DDT as a contact poison against several arthropods" in 1948.

By October 1945, DDT was available for public sale in the United States. Although it was promoted by government and industry for use as an agricultural and household pesticide, there were also concerns about its use from the beginning. Opposition to DDT was focused by the 1962 publication of Rachel Carson's book *Silent Spring*. It cataloged environmental impacts that coincided with widespread use of DDT in agriculture in the United States, and it questioned the logic of broadcasting potentially dangerous chemicals into the environment with little prior investigation of their environmental and health effects. The book claimed that DDT and other pesticides had been shown to cause cancer and that their agricultural use was a threat to wildlife, particularly birds. Its publication was a seminal event for the environmental movement and resulted in a large public outcry that eventually led, in 1972, to a ban on DDT's agricultural use in the United States. A worldwide ban on agricultural use was formalized under the Stockholm Convention on Persistent Organic Pollutants, but its limited and still-controversial use in disease vector control continues, because of its effectiveness in reducing malarial infections, balanced by environmental and other health concerns.

Along with the passage of the Endangered Species Act, the United States ban on DDT is a major factor in the comeback of the bald eagle (the national bird of the United States) and the peregrine falcon from near-extinction in the contiguous United States.

Properties and Chemistry

DDT is similar in structure to the insecticide methoxychlor and the acaricide dicofol. It is highly hydrophobic and nearly insoluble in water but has good solubility in most organic solvents, fats and oils. DDT does not occur naturally and is synthesised by consecutive Friedel–Crafts reactions between chloral (CCl_3CHO) and two equivalents of chlorobenzene (C_6H_5Cl), in the presence of an acidic catalyst. DDT has been marketed under trade names including Anofex, Cezarex, Chlorophenothane, Dicophane, Dinocide, Gesarol, Guesapon, Guesarol, Gyron, Ixodex, Neocid, Neocidol and Zerdane; INN is clofenotane.

Isomers and Related Compounds

Commercial DDT is a mixture of several closely–related compounds. Due to the nature of the chemical reaction used to synthesize DDT, several combinations of *ortho* and *para* arene substitution patterns are formed. The major component (77%) is the desired *p,p'* isomer. The *o,p'* isomeric impurity is also present in significant amounts (15%). Dichlorodiphenyldichloroethylene (DDE) and dichlorodiphenyldichloroethane (DDD) make up the balance of impurities in commercial samples. DDE and DDD are also the major metabolites and environmental breakdown products. DDT, DDE and DDD are sometimes referred to collectively as DDX.

Components of commercial DDT.

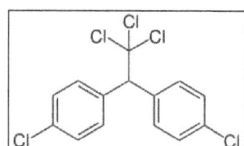

p,p'-DDT
(desired compound)

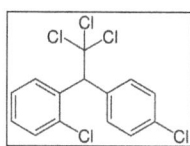

o,p'-DDT
(isomeric impurity)

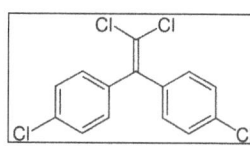

p,p'-DDE
(impurity)

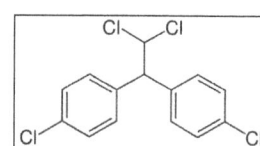

p,p'-DDD
(impurity)

Production and Use

DDT has been formulated in multiple forms, including solutions in xylene or petroleum distillates, emulsifiable concentrates, water-wettable powders, granules, aerosols, smoke candles and charges for vaporizers and lotions.

From 1950 to 1980, DDT was extensively used in agriculture – more than 40,000 tonnes each year worldwide – and it has been estimated that a total of 1.8 million tonnes have been produced globally since the 1940s. In the United States, it was manufactured by some 15 companies, including Monsanto, Ciba, Montrose Chemical Company, Pennwalt, and Velsicol Chemical Corporation. Production peaked in 1963 at 82,000 tonnes per year. More than 600,000 tonnes (1.35 billion pounds) were applied in the US before the 1972 ban. Usage peaked in 1959 at about 36,000 tonnes.

In 2009, 3,314 tonnes were produced for malaria control and visceral leishmaniasis. India is the only country still manufacturing DDT, and is the largest consumer. China ceased production in 2007.

Mechanism of Insecticide Action

In insects, DDT opens sodium ion channels in neurons, causing them to fire spontaneously, which leads to spasms and eventual death. Insects with certain mutations in their sodium channel gene are resistant to DDT and similar insecticides. DDT resistance is also conferred by up-regulation of genes expressing cytochrome P450 in some insect species, as greater quantities of some enzymes of this group accelerate the toxin's metabolism into inactive metabolites. (The same enzyme family is up-regulated in mammals too, e.g., in response to ethanol consumption.) Genomic studies in the model genetic organism *Drosophila melanogaster* revealed that high level DDT resistance is polygenic, involving multiple resistance mechanisms.

Environmental Impact

DDT is a persistent organic pollutant that is readily adsorbed to soils and sediments, which can act both as sinks and as long-term sources of exposure affecting organisms. Depending on conditions, its soil half-life can range from 22 days to 30 years. Routes of loss and degradation include runoff, volatilization, photolysis and aerobic and anaerobic biodegradation. Due to hydrophobic properties, in aquatic ecosystems DDT and its metabolites are absorbed by aquatic organisms and adsorbed on suspended particles, leaving little DDT dissolved in the water (however, its half-life in aquatic environments is listed by the National Pesticide Information Center as 150 years). Its breakdown products and metabolites, DDE and DDD, are also persistent and have similar chemical and physical properties. DDT and its breakdown products are transported from warmer areas to the Arctic by the phenomenon of global distillation, where they then accumulate in the region's food web.

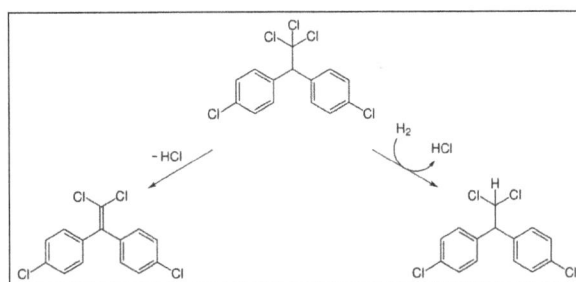

Degradation of DDT to form DDE (by elimination of HCl, left) and DDD (by reductive dechlorination, right).

Medical researchers in 1974 found a measurable and significant difference in the presence of DDT in human milk between mothers who lived in New Brunswick and mothers who lived in Nova Scotia, "possibly because of the wider use of insecticide sprays in the past".

Because of its lipophilic properties, DDT can bioaccumulate, especially in predatory birds. DDT is toxic to a wide range of living organisms, including marine animals such as crayfish, daphnids, sea shrimp and many species of fish. DDT, DDE and DDD magnify through the food chain, with apex predators such as raptor birds concentrating more chemicals than other animals in the same environment. They are stored mainly in body fat. DDT and DDE are resistant to metabolism; in humans, their half-lives are 6 and up to 10 years, respectively. In the United States, these chemicals were detected in almost all human blood samples tested by the Centers for Disease Control in 2005, though their levels have sharply declined since most uses were banned. Estimated dietary intake has declined, although FDA food tests commonly detect it.

Despite being banned for many years, in 2018 research showed that DDT residues are still present in European soils and Spanish rivers.

Eggshell Thinning

The chemical and its breakdown products DDE and DDD caused eggshell thinning and population declines in multiple North American and European bird of prey species. DDE-related eggshell thinning is considered a major reason for the decline of the bald eagle, brown pelican, peregrine falcon and osprey. However, birds vary in their sensitivity to these chemicals, with birds of prey, waterfowl and song birds being more susceptible than chickens and related species. Even in 2010, California condors that feed on sea lions at Big Sur that in turn feed in the Palos Verdes Shelf area of the Montrose Chemical Superfund site exhibited continued thin-shell problems, though DDT's role in the decline of the California condor is disputed.

The biological thinning mechanism is not entirely understood, but DDE appears to be more potent than DDT, and strong evidence indicates that p,p'-DDE inhibits calcium ATPase in the membrane of the shell gland and reduces the transport of calcium carbonate from blood into the eggshell gland. This results in a dose-dependent thickness reduction. Other evidence indicates that o,p'-DDT disrupts female reproductive tract development, later impairing eggshell quality. Multiple mechanisms may be at work, or different mechanisms may operate in different species.

Human Health

A U.S. soldier is demonstrating DDT hand-spraying equipment. DDT was used to control the spread of typhus-carrying lice.

Spraying hospital beds with DDT, PAIGC hospital of Ziguinchor, 1973.

DDT is an endocrine disruptor. It is considered likely to be a human carcinogen although the majority of studies suggest it is not directly genotoxic. DDE acts as a weak androgen receptor antagonist, but not as an estrogen. p,p'-DDT, DDT's main component, has little or no androgenic or estrogenic activity. The minor component o,p'-DDT has weak estrogenic activity.

Acute Toxicity

DDT is classified as "moderately toxic" by the US National Toxicology Program (NTP) and "moderately hazardous" by WHO, based on the rat oral LD_{50} of 113 mg/kg. Indirect exposure is considered relatively non-toxic for humans.

Chronic Toxicity

Primarily through the tendency for DDT to buildup in areas of the body with high lipid content, chronic exposure can affect reproductive capabilities and the embryo or fetus.

- In The Lancet States, "research has shown that exposure to DDT at amounts that would be needed in malaria control might cause preterm birth and early weaning toxicological evidence shows endocrine-disrupting properties; human data also indicate possible disruption in semen quality, menstruation, gestational length, and du-ration of lactation."

- Other studies document decreases in semen quality among men with high exposures (generally from IRS).

- Studies are inconsistent on whether high blood DDT or DDE levels increase time to pregnancy. In mothers with high DDE blood serum levels, daughters may have up to a 32% increase in the probability of conceiving, but increased DDT levels have been associated with a 16% decrease in one study.

- Indirect exposure of mothers through workers directly in contact with DDT is associated with an increase in spontaneous abortions.

- Other studies found that DDT or DDE interfere with proper thyroid function in pregnancy and childhood.

- Mothers with high levels of DDT circulating in their blood during pregnancy were found to be more likely to give birth to children who would go on to develop autism.

Carcinogenicity

In 2015, the International Agency for Research on Cancer classified DDT as Group 2A "probably carcinogenic to humans". Previous assessments by the U.S. National Toxicology Program classified it as "reasonably anticipated to be a carcinogen" and by the EPA classified DDT, DDE and DDD as class B2 "probable" carcinogens; these evaluations were based mainly on animal studies.

A 2005 Lancet review stated that occupational DDT exposure was associated with increased pancreatic cancer risk in 2 case control studies, but another study showed no DDE dose-effect association. Results regarding a possible association with liver cancer and biliary tract cancer are conflicting: workers who did not have direct occupational DDT contact showed increased risk. White men had an increased risk, but not white women or black men. Results about an association with multiple myeloma, prostate and testicular cancer, endometrial cancer and colorectal cancer have been inconclusive or generally do not support an association. A 2017 review of liver cancer studies concluded that "organochlorine pesticides, including DDT, may increase hepatocellular carcinoma risk".

A 2009 review, whose co-authors included persons engaged in DDT-related litigation, reached broadly similar conclusions, with an equivocal association with testicular cancer. Case–control studies did not support an association with leukemia or lymphoma.

Breast Cancer

The question of whether DDT or DDE are risk factors in breast cancer has not been conclusively answered. Several meta analyses of observational studies have concluded that there is no overall relationship between DDT exposure and breast cancer risk. The United States Institute of Medicine reviewed data on the association of breast cancer with DDT exposure in 2012 and concluded that a causative relationship could neither be proven nor disproven.

A 2007 case–control study using archived blood samples found that breast cancer risk was increased 5-fold among women who were born prior to 1931 and who had high serum DDT levels in 1963. Reasoning that DDT use became widespread in 1945 and peaked around 1950, they concluded that the ages of 14–20 were a critical period in which DDT exposure leads to increased risk. This study, which suggests a connection between DDT exposure and breast cancer that would not be picked up by most studies, has received variable commentary in third party reviews. One review suggested that "previous studies that measured exposure in older women may have missed the critical period". The National Toxicology Program notes that while the majority of studies have not found a relationship between DDT exposure and breast cancer that positive associations have been seen in a "few studies among women with higher levels of exposure and among certain subgroups of women".

A 2015 case control study identified a link (odds ratio 3.4) between *in-utero* exposure (as estimated from archived maternal blood samples) and breast cancer diagnosis in daughters. The findings "support classification of DDT as an endocrine disruptor, a predictor of breast cancer, and a marker of high risk".

Malaria Control

Malaria remains the primary public health challenge in many countries. In 2015, there were 214 million cases of malaria worldwide resulting in an estimated 438,000 deaths, 90% of which occurred in Africa. DDT is one of many tools to fight the disease. Its use in this context has been called everything from a "miracle weapon that is like Kryptonite to the mosquitoes", to "toxic colonialism".

Before DDT, eliminating mosquito breeding grounds by drainage or poisoning with Paris green or pyrethrum was sometimes successful. In parts of the world with rising living standards, the elimination of malaria was often a collateral benefit of the introduction of window screens and improved sanitation. A variety of usually simultaneous interventions represents best practice. These include antimalarial drugs to prevent or treat infection; improvements in public health infrastructure to diagnose, sequester and treat infected individuals; bednets and other methods intended to keep mosquitoes from biting humans; and vector control strategies such as larvaciding with insecticides, ecological controls such as draining mosquito breeding grounds or introducing fish to eat larvae and indoor residual spraying (IRS) with insecticides, possibly including DDT. IRS involves the treatment of interior walls and ceilings with insecticides. It is particularly effective against mosquitoes, since many species rest on an indoor wall before or after feeding. DDT is one of 12 WHO–approved IRS insecticides.

The WHO's anti-malaria campaign of the 1950s and 1960s relied heavily on DDT and the results were promising, though temporary in developing countries. Experts tie malarial resurgence to multiple factors, including poor leadership, management and funding of malaria control programs; poverty; civil unrest; and increased irrigation. The evolution of resistance to first-generation drugs (e.g. chloroquine) and to insecticides exacerbated the situation. Resistance was largely fueled by unrestricted agricultural use. Resistance and the harm both to humans and the environment led many governments to curtail DDT use in vector control and agriculture. In 2006 WHO reversed a longstanding policy against DDT by recommending that it be used as an indoor pesticide in regions where malaria is a major problem.

Once the mainstay of anti-malaria campaigns, as of 2008 only 12 countries used DDT, including India and some southern African states, though the number was expected to rise.

Initial Effectiveness

When it was introduced in World War II, DDT was effective in reducing malaria morbidity and mortality. WHO's anti-malaria campaign, which consisted mostly of spraying DDT and rapid treatment and diagnosis to break the transmission cycle, was initially successful as well. For example, in Sri Lanka, the program reduced cases from about one million per year before spraying to just 18 in 1963 and 29 in 1964. Thereafter the program was halted to save money and malaria rebounded to 600,000 cases in 1968 and the first quarter of 1969. The country resumed DDT vector control but the mosquitoes had evolved resistance in the interim, presumably because of continued agricultural use. The program switched to malathion, but despite initial successes, malaria continued its resurgence into the 1980s.

DDT remains on WHO's list of insecticides recommended for IRS. After the appointment of Arata Kochi as head of its anti-malaria division, WHO's policy shifted from recommending IRS only in areas of seasonal or episodic transmission of malaria, to advocating it in areas of continuous, intense transmission. WHO reaffirmed its commitment to phasing out DDT, aiming "to achieve a 30% cut in the application of DDT world-wide by 2014 and its total phase-out by the early 2020s if not sooner" while simultaneously combating malaria. WHO plans to implement alternatives to DDT to achieve this goal.

South Africa continues to use DDT under WHO guidelines. In 1996, the country switched to alternative insecticides and malaria incidence increased dramatically. Returning to DDT and introducing new drugs brought malaria back under control. Malaria cases increased in South America after countries in that continent stopped using DDT. Research data showed a strong negative relationship between DDT residual house sprayings and malaria. In a research from 1993 to 1995, Ecuador increased its use of DDT and achieved a 61% reduction in malaria rates, while each of the other countries that gradually decreased its DDT use had large increases.

Mosquito Resistance

In some areas resistance reduced DDT's effectiveness. WHO guidelines require that absence of resistance must be confirmed before using the chemical. Resistance is largely due to agricultural use, in much greater quantities than required for disease prevention.

Resistance was noted early in spray campaigns. Paul Russell, former head of the Allied Anti-Malaria campaign, observed in 1956 that "resistance has appeared after six or seven years". Resistance has been detected in Sri Lanka, Pakistan, Turkey and Central America and it has largely been replaced by organophosphate or carbamate insecticides, e.g. malathion or bendiocarb.

In many parts of India, DDT is ineffective. Agricultural uses were banned in 1989 and its anti-malarial use has been declining. Urban use ended. One study concluded that "DDT is still a viable insecticide in indoor residual spraying owing to its effectivity in well supervised spray operation and high excito-repellency factor."

Studies of malaria-vector mosquitoes in KwaZulu-Natal Province, South Africa found susceptibility to 4% DDT (WHO's susceptibility standard), in 63% of the samples, compared to the average of 87% in the same species caught in the open. The authors concluded that "Finding DDT resistance in the vector *An. arabiensis*, close to the area where we previously reported pyrethroid-resistance in the vector *An. funestus* Giles, indicates an urgent need to develop a strategy of insecticide resistance management for the malaria control programmes of southern Africa."

DDT can still be effective against resistant mosquitoes and the avoidance of DDT-sprayed walls by mosquitoes is an additional benefit of the chemical. For example, a 2007 study reported that resistant mosquitoes avoided treated huts. The researchers argued that DDT was the best pesticide for use in IRS (even though it did not afford the most protection from mosquitoes out of the three test chemicals) because the other pesticides worked primarily by killing or irritating mosquitoes – encouraging the development of resistance. Others argue that the avoidance behavior slows eradication. Unlike other insecticides such as pyrethroids, DDT requires long exposure to accumulate a lethal dose; however its irritant property shortens contact periods. "For these reasons, when comparisons have been made, better malaria control has generally been achieved with pyrethroids than with DDT." In India outdoor sleeping and night duties are common, implying that "the excito-repellent effect of DDT, often reported useful in other countries, actually promotes outdoor transmission".

Residents' Concerns

IRS is effective if at least 80% of homes and barns in a residential area are sprayed. Lower coverage rates can jeopardize program effectiveness. Many residents resist DDT spraying, objecting to the lingering smell, stains on walls, and the potential exacerbation of problems with other insect pests. Pyrethroid insecticides (e.g. deltamethrin and lambda-cyhalothrin) can overcome some of these issues, increasing participation.

Human Exposure

A 1994 study found that South Africans living in sprayed homes have levels that are several orders of magnitude greater than others. Breast milk from South African mothers contains high levels of DDT and DDE. It is unclear to what extent these levels arise from home spraying vs food residues. Evidence indicates that these levels are associated with infant neurological abnormalities.

Most studies of DDT's human health effects have been conducted in developed countries where DDT is not used and exposure is relatively low.

Illegal diversion to agriculture is also a concern as it is difficult to prevent and its subsequent use on crops is uncontrolled. For example, DDT use is widespread in Indian agriculture, particularly mango production and is reportedly used by librarians to protect books. Other examples include Ethiopia, where DDT intended for malaria control is reportedly used in coffee production, and Ghana where it is used for fishing. The residues in crops at levels unacceptable for export have been an important factor in bans in several tropical countries. Adding to this problem is a lack of skilled personnel and management.

Criticism of Restrictions on DDT Use

A few people and groups have argued that limitations on DDT use for public health purposes have caused unnecessary morbidity and mortality from vector-borne diseases, with some claims of malaria deaths ranging as high as the hundreds of thousands and millions. Robert Gwadz of the US National Institutes of Health said in 2007, "The ban on DDT may have killed 20 million children." These arguments were rejected as "outrageous" by former WHO scientist Socrates Litsios. May Berenbaum, University of Illinois entomologist, says, "to blame environmentalists who oppose DDT for more deaths than Hitler is worse than irresponsible". More recently, Dr. Michael Palmer, a professor of chemistry at the University of Waterloo, has pointed out that DDT is still used to prevent malaria, that its declining use is primarily due to increases in manufacturing costs, and that in Africa, efforts to control malaria have been regional or local, not comprehensive.

Criticisms of a DDT "ban" often specifically reference the 1972 United States ban (with the erroneous implication that this constituted a worldwide ban and prohibited use of DDT in vector control). Reference is often made to *Silent Spring*, even though Carson never pushed for a DDT ban. John Quiggin and Tim Lambert wrote, "the most striking feature of the claim against Carson is the ease with which it can be refuted".

Alternatives

Insecticides

Organophosphate and carbamate insecticides, e.g. malathion and bendiocarb, respectively, are more expensive than DDT per kilogram and are applied at roughly the same dosage. Pyrethroids such as deltamethrin are also more expensive than DDT, but are applied more sparingly (0.02–0.3 g/m^2 vs 1–2 g/m^2), so the net cost per house per treatment is about the same. It must be noted that DDT has one of the longest resiudal efficacy periods of any IRS insecticide, lasting 6 to 12 months. Pyrethroids will remain active for only 4 to 6 months, and organophosphates and carbamates remain active for 2 to 6 months. In many malaria-endemic countries, malaria transmission occurs year-round, meaning that the high expense conducting a spray campaign (including hiring spray operators, procuring insecticides, and conducting pre-spray outreach campaings to encourage people to be home and to accept the intervention) will need to occur multiple times per year for these shorter lasting insecticides.

Non-chemical Vector Control

Before DDT, malaria was successfully eliminated or curtailed in several tropical areas by removing or poisoning mosquito breeding grounds and larva habitats, for example by eliminating standing water. These methods have seen little application in Africa for more than half a century. According to CDC, such methods are not practical in Africa because "*Anopheles gambiae*, one of the primary vectors of malaria in Africa, breeds in numerous small pools of water that form due to rainfall. It is difficult, if not impossible, to predict when and where the breeding sites will form, and to find and treat them before the adults emerge."

The relative effectiveness of IRS versus other malaria control techniques (e.g. bednets or prompt access to anti-malarial drugs) varies and is dependent on local conditions.

A WHO study released in January 2008 found that mass distribution of insecticide-treated mosquito nets and artemisinin–based drugs cut malaria deaths in half in malaria-burdened Rwanda and Ethiopia. IRS with DDT did not play an important role in mortality reduction in these countries.

Vietnam has enjoyed declining malaria cases and a 97% mortality reduction after switching in 1991 from a poorly funded DDT-based campaign to a program based on prompt treatment, bednets and pyrethroid group insecticides.

In Mexico, effective and affordable chemical and non-chemical strategies were so successful that the Mexican DDT manufacturing plant ceased production due to lack of demand.

A review of fourteen studies in sub-Saharan Africa, covering insecticide-treated nets, residual spraying, chemoprophylaxis for children, chemoprophylaxis or intermittent treatment for pregnant women, a hypothetical vaccine and changing front–line drug treatment, found decision making limited by the lack of information on the costs and effects of many interventions, the small number of cost-effectiveness analyses, the lack of evidence on the costs and effects of packages of measures and the problems in generalizing or comparing studies that relate to specific settings and use different methodologies and outcome measures. The two cost-effectiveness estimates of DDT residual spraying examined were not found to provide an accurate estimate of the cost-effectiveness of DDT spraying; the resulting estimates may not be good predictors of cost-effectiveness in current programs.

However, a study in Thailand found the cost per malaria case prevented of DDT spraying (US$1.87) to be 21% greater than the cost per case prevented of lambda-cyhalothrin–treated nets (US$1.54), casting some doubt on the assumption that DDT was the most cost-effective measure. The director of Mexico's malaria control program found similar results, declaring that it was 25% cheaper for Mexico to spray a house with synthetic pyrethroids than with DDT. However, another study in South Africa found generally lower costs for DDT spraying than for impregnated nets.

A more comprehensive approach to measuring cost-effectiveness or efficacy of malarial control would not only measure the cost in dollars, as well as the number of people saved, but would also consider ecological damage and negative human health impacts. One preliminary study found that it is likely that the detriment to human health approaches or exceeds the beneficial reductions in malarial cases, except perhaps in epidemics. It is similar to the earlier study regarding estimated theoretical infant mortality caused by DDT and subject to the criticism.

A study in the Solomon Islands found that "although impregnated bed nets cannot entirely replace DDT spraying without substantial increase in incidence, their use permits reduced DDT spraying".

A comparison of four successful programs against malaria in Brazil, India, Eritrea and Vietnam does not endorse any single strategy but instead states, "Common success factors included conducive country conditions, a targeted technical approach using a package of effective tools, data-driven decision-making, active leadership at all levels of government, involvement of communities, decentralized implementation and control of finances, skilled technical and managerial capacity at national and sub-national levels, hands-on technical and programmatic support from partner agencies, and sufficient and flexible financing."

DDT resistant mosquitoes may be susceptible to pyrethroids in some countries. However, pyrethroid resistance in *Anopheles* mosquitoes is on the rise with resistant mosquitoes found in multiple countries.

DIFLUBENZURON

Diflubenzuron is an insecticide of the benzoylurea class. It is used in forest management and on field crops to selectively control insect pests, particularly forest tent caterpillar moths, boll weevils, gypsy moths, and other types of moths. It is a widely used larvicide in India for control of mosquito larvae by public health authorities. Diflubenzuron is approved by the WHO Pesticide Evaluation Scheme.

Mechanism of Action

The mechanism of action of diflubenzuron involves inhibiting the production of chitin which is used by an insect to build its exoskeleton. It triggers insect larvae to molt early without a properly formed exoskeleton, resulting in the death of the larvae.

Environmental Toxicity

Diflubenzuron has been evaluated by the United States Environmental Protection Agency (EPA), and it is classified as non-carcinogenic. 4-Chloroaniline, a metabolite of diflubenzuron which has been classified as a carcinogen, is produced after diflubenzuron has been ingested. The small amount converted to 4-chloroaniline after ingestion is not sufficient to cause cancer.

Commercial Uses

A commercial preparation containing diflubenzuron is sold under the trade name Adept and is used as an insect growth regulator designed to kill fungus gnat larvae in commercial greenhouses. It is applied to infected soil and will kill fungus gnat larvae for 30-60 days from a single application. Although it is targeted at fungus gnat larvae, care should be taken in applying it as it is highly toxic to most aquatic invertebrates. It has no toxic effects on adult insects, only insect larvae are affected. Diflubenzuron can cause serious foliar injury to plants in the spurge family and certain types of begonia, particularly poinsettias, hibiscus and reiger begonia and should not be applied to these plant varieties.

Diflubenzuron is used as a larvicide in the cattle farming industry. Sold under the name Vigilante, it is formulated as a bolus and is used to control fly populations.

INSECT GROWTH REGULATORS

An insect growth regulator (IGR) is a substance (chemical) that inhibits the life cycle of an insect. IGRs are typically used as insecticides to control populations of harmful insect pests such as cockroaches and fleas.

Advantages

Many IGRs are labeled "reduced risk" by the Environmental Protection Agency, meaning that they target juvenile harmful insect populations while causing less detrimental effects to beneficial insects. Many beekeepers have reported IGR's negatively affecting brood and young bees . Unlike classic insecticides, IGRs do not affect an insect's nervous system and are thus more friendly to "worker insects" within closed environments. IGRs are also more compatible with pest management systems that use biological controls. In addition, while insects can become resistant to insecticides, they are less likely to become resistant to IGRs.

Mechanism of Action

As an insect grows it molts, growing a new exoskeleton under its old one and then shedding the old one to allow the new one to swell to a new size and harden. IGRs prevent an insect from reaching maturity by interfering with the molting process. This in turn curbs infestations because immature insects cannot reproduce. Because these IGRs work by interfering with an insect's molting process, they kill insects more slowly than traditional insecticides. Death typically occurs within 3 to 10 days, depending on the IGR product, the insect's life stage at the time the product is applied, and how quickly the insect develops. Some IGRs cause insects to stop feeding long before they die.

Hormonal IGRs

Hormonal IGRs typically work by mimicking or inhibiting the juvenile hormone (JH), one of the two major hormones involved in insect molting. IGRs can also inhibit the other hormone, ecdysone, large peaks of which trigger the insect to molt. If JH is present at the time of molting, the insect molts into a larger larval form; if absent, it molts into a pupa or adult. IGRs that mimic JH can produce premature molting of young immature stages, disrupting larval development. They can also act on eggs, causing sterility, disrupting behavior or disrupting diapause, the process that causes an insect to become dormant before winter. IGRs that inhibit JH production can cause insects to prematurely molt into a nonfunctional adult. IGRs that inhibit ecdysone can cause pupal mortality by interrupting the transformation of larval tissues into adult tissues during the pupal stage.

Chitin Synthesis Inhibitors

Chitin synthesis inhibitors work by preventing the formation of chitin, a carbohydrate needed to form the insect's exoskeleton. With these inhibitors, an insect grows normally until it molts. The inhibitors prevent the new exoskeleton from forming properly, causing the insect to die. Death may be quick, or take up to several days depending on the insect. Chitin synthesis inhibitors can also kill eggs by disrupting normal embryonic development. Chitin synthesis inhibitors affect insects for longer periods of time than hormonal IGRs. These are also quicker acting but can affect predaceous insects, arthropods and even fish. Compounds include benzoylurea pesticides.

Examples:

- Diflubenzuron (Vigilante),
- Azadirachtin (AzaGuard),
- Hydroprene (Gentrol),
- Methoprene (Precor),
- Pyriproxyfen (Nyguard, Nylar, Sumilarv),
- Triflumuron (Starycide).

Methoprene

Methoprene, an insect growth regulator, is being applied to many home and community pest control problems as a general use, slow-acting insecticide. This chemical can be used to control a number of pests, including fleas (PrecorTM), mosquitoes (AltosidTM), pharaoh ants, leaf miners and hoppers, and cucmber beetles. It is also used as an insect control in food production and agriculture. EPA estimated in 1982 that 57% of use at that time was as an additive to cattle feed and mineral supplements to control horn flies.

Mode of action

This chemical is an analog to a unique insect-growth regulating hormone, which does not ressemble any known mammalian hormones. Use requires careful attention to timing and patience. Applied at very low rates, while insect populations are still in the egg or larval stage of their life cycle, methoprene prevents development to the adult reproductive stages so that insects die in arrested immaturity. Methoprene is not toxic when applied to adult stages of the target insect. Because the chemical interferes with the insect's normal life cycle and is not directly toxic to the pest, it is considered to be a biochemical pesticide.

Toxicity

According to information contained in a 1982 EPA Registration Standard, methoprene is of extremely low acute toxicity to mammals (LD50 equals 36,500 mg/kg). It is not a skin or eye irritant, although it is slightly toxic via dermal absorbtion. For this effect, methoprene is a toxicity category III with the signal word CAUTION required on the label.

No adverse effects have been reported in animal bio-assays for long-term health effects and short-term tests for mutagenicity were all negative. EPA reviewers found that animals rapidly metabolize and excrete the material, the major nonwater soluble metabolite in animal assays being cholesterol.

Ecological Effects

Methoprene may have severe developmental effects on frogs. It was found to be a possible cause of a sharp rise in the incidence of frog deformities throughout North America. A 1997 study linked pesticides to frog deformities when they found a higher number of hindlimb frog deformities occurring in agricultural areas.

It is believed that a breakdown product of methoprene mimics retinoic acid, an important chemical to the development of fish and frog embryos. Laboratory tests involving raised levels of retinoic acid have resulted in a majority of the limb deformities found in the North American frogs.

Methoprene also has a moderate toxicity towards both warm and cold water, freshwater fish, although exposure of these organisms is limited due to methoprene's rapid degradation in unshaded water. It is highly acutely toxic to estuarine and marine invertebrates, which play an important role in the delicate estuarine ecosystem. The LC50 (concentration needed to kill half of the test population) for fresh water shrimp is greater than 0.1 ppb (parts per billion) and the LC50 for the estuarine mud crabs is greater than .0001 ppb. Meanwhile, the level of methoprene released into an environment from a general application is expected to be around 10 ppb.

Environmental Fate

Studies reviewed by EPA indicate that if protected from light, methoprene is quite stable in water within the pH range 5-9, not degrading after 30 days in the dark. Methoprene used inside homes is active for at least 6 months against developing fleas. When exposed to light, however, methoprene degrades within 7 days to more than 50 products, not all of which have been identified.

The chemical's soil half-life is between 10-14 days in four soils tested, where it is microbially degraded to carbon dioxide and soil-bound products.

Although the mode of action and low persistence of methoprene imply that resistance problems should be slow to develop, resistance to insect growth hormones has been induced experimentally and therefore might be possible in the field.

Hydroprene

Hydroprene is an insecticide used against cockroaches, beetles, and moths. It was registered by the U.S. Environmental Protection Agency (EPA) in 1984. Hydroprene belongs to the class of insecticides known as insect growth regulators (IGRs). The U.S. EPA currently categorizes hydroprene as a biopesticide based on its biochemical properties as an IGR. Hydroprene is an yellowish-brown liquid that is soluble in water and slightly volatile. It is stable for more than 3 years under normal storage conditions. Signal words for hydroprene products range from Caution to Warning. The signal word reflects the combined toxicity of hydroprene and other product ingredients. Hydroprene products are used on a variety of indoor sites including homes, offices, warehouses, restaurants, hospitals, and greenhouses. Commercial formulations of the insecticide include aerosols, liquids, and impregnated materials (i.e., bait stations).

Working of Hydroprene

Hydroprene disrupts normal development and molting of insects by mimicking hormones produced by immature insects. Hydroprene causes different effects on different insects. It may cause adult sterility, physical body changes, water loss, and premature death.

Some Products that Contain Hydroprene

- Gencor,
- Gentrol,
- Raid Max Sterilizer Discs.

Toxicity of Hydroprene

Animals: Hydroprene is very low in toxicity when ingested by rats and dogs. Hydroprene is low in toxicity when applied to the skin of rabbits and rats. Hydroprene is very low in toxicity when inhaled by rats. In a skin irritation studies, hydroprene was non-irritating to rabbits and mildly irritating to rats. The U.S. EPA classifies hydroprene as very low in toxicity for skin effects. In an eye irritation study with rabbits, hydroprene caused mild eye irritation. The U.S. EPA classifies hydroprene as very low in toxicity for eye effects. The U.S. EPA classifies hydroprene as very low in its ability to increase the sensitivity of the skin to chemical exposure. Scientists fed rats hydroprene for 28 days and noted kidney effects at high doses. In a 90-day feeding study, investigators exposed male and female rats to hydroprene. At high doses, scientists observed liver cell effects in both sexes and ovarian cell effects in female rats. No effects occurred at low doses.

Humans: Data are not available from occupational exposure, accidental poisonings, or epidemiological studies regarding the acute toxicity of hydroprene.

Does Hydroprene Break Down and Leave the Body?

Animals: In rats fed hydroprene, levels of the chemical peaked in the blood at 5 to 7 hours. Researchers detected the highest levels in the liver, fat, and adrenal glands after 6 hours. Elimination from the body began on the day of dosing and continued for several days. Approximately 13% of the hydroprene was retained in rats.

Humans: Data are not available regarding the break down and elimination of hydroprene in humans.

Does Hydroprene Cause Reproductive or Birth Defects?

Animals: In a reproductive study, researchers fed rats hydroprene in the diet. They detected no fertility effects. Adult and offspring rats fed the highest doses had lower body weight gains. In a developmental study, scientists exposed pregnant rabbits to hydroprene and detected no developmental effects. At the highest dose, scientist noted maternal weight loss.

Humans: Data are not available from work-related exposure, accidental poisonings, or other human studies regarding the reproductive and developmental toxicity of hydroprene.

Does Hydroprene Cause Cancer?

Animals: In a cancer study, laboratory workers fed rats hydroprene in their diets. Workers noted no evidence of cancer. Methoprene, an IGR that is similar to hydroprene, has not displayed an ability to cause cancer. Researchers often test chemicals for their ability to change the genetic material of an organism as an indication of the chemical's potential to cause cancer. Sufficient evidence exists to determine that hydroprene does not have significant potential to change genetic material.

Humans: The U.S. EPA currently classifies hydroprene as a group D carcinogen. This classification means that not enough data exist to say whether hydroprene causes cancer in humans. Data are not available from work-related exposures or other human studies regarding the potential of hydroprene to cause cancer.

Environmental Fate and Behavior of Hydroprene

Hydroprene is rapidly degraded in soil with a half-life of a few days. No data are available regarding the fate of hydroprene in water. Hydroprene is degraded by plants.

Effects of Hydroprene on Wildlife

Hydroprene is practically non-toxic to fish. It may be toxic to other water organisms by affecting their development. No data are available regarding the toxicity of hydroprene to birds. Hydroprene is low in toxicity to adult bees. It may affect immature bees.

References

- Metcalf, Robert L. (2002). "Insect Control". Ullmann's Encyclopedia of Industrial Chemistry. Wiley-VCH. Doi:10.1002/14356007.a14_263. ISBN 978-3527306732

- Insecticide, insecticide: britannica.com, Retrieved 30 January, 2019

- IUPAC (2006). "Glossary of Terms Relating to Pesticides"(PDF). IUPAC. P. 2123. Retrieved January 28, 2014

- Ware-intro-insecticides: umn.edu, Retrieved 1 February, 2019

- Junquera, Pablo; Hosking, Barry; Gameiro, Marta; Macdonald, Alicia (2019). "Benzoylphenyl ureas as veterinary antiparasitics. An overview and outlook with emphasis on efficacy, usage and resistance". Parasite. 26: 26. Doi:10.1051/parasite/2019026. ISSN 1776-1042. PMID 31041897

- Methoprene, factsheets, pesticides, documents, media, assets: beyondpesticides.org, Retrieved 3 March, 2019

- Kydonieus, Agis F. (2017-10-02). Treatise on Controlled Drug Delivery: Fundamentals-optimization-applications. Routledge. ISBN 9781351406871

- Hydropregen, factsheets: orst.edu, Retrieved 4 April, 2019

- "igrs -- A Growing, But Misunderstood Group". GPN: Greenhouse Product News. Archived from the original on 2011-07-11. Retrieved 20 November 2010

4

Organochlorine Pesticides and Insecticides

Organochlorine pesticides and insecticides are chlorinated hydrocarbons which are widely used all over the world. It includes DDT, methoxychlor, dieldrin, chlordane, toxaphene, mirex, kepone, lindane, and benzene hexachloride. The topics elaborated in this chapter will help in gaining a better perspective about organochlorine pesticides and insecticides.

Organochlorine (OC) pesticides are synthetic pesticides widely used all over the world. They belong to the group of chlorinated hydrocarbon derivatives, which have vast application in the chemical industry and in agriculture. These compounds are known for their high toxicity, slow degradation and bioaccumulation. Even though many of the compounds which belong to OC were banned in developed countries, the use of these agents has been rising. This concerns particularly abuse of these chemicals which is in practice across the continents. Though pesticides have been developed with the concept of target organism toxicity, often non-target species are affected badly by their application.

Organochlorine Pesticides

The basic characteristics of organochlorine pesticides are high persistence, low polarity, low aqueous solubility and high lipid solubility. Organochlorine pesticides can enter the environment after pesticide applications, polluted wastes discarded into landfills, and discharges from industrial units that synthesize these chemicals. They are volatile and stable; some can adhere to the soil and air, thus increasing the chances of high persistence in the environment, and are identified as agents of chronic exposure to animals and humans.

They have a related chemical structure, showing chlorine substituted aliphatic or aromatic rings. Due to their structural resemblances, these compounds share certain physicochemical characteristics such as persistence, bioaccumulation and toxicity. One basic character that they share across the spectrum is persistence, where persistence is defined as half-life greater than two months in water or six months in soil sediment. The persistence of OC compounds varies from moderate persistence with half-life of approximately 60 days to high persistence with half-life up to 10–15 years. The most commonly used pesticide in agricultural practice is dichlorodiphenyltrichloroethane (DDT), which is moderately hazardous, with high persistence and a half-life of 2–15 years. The use of DDT is now banned in many countries but it is illegally used in most of the developing countries. This applies also to endosulphan, an insecticide which is highly hazardous and has moderate persistence with a half-life of fifty days and is used in the production of cashew.

Due to the high persistence and bioaccumulation potential, the Stockholm Convention has classified most of the OC compounds as environmental hazards and banned the use of many of them. However in many developing countries they are still in use making the ban ineffective.

Biochemical Toxicity of Organochlorines

Organochlorine toxicity is mainly due to stimulation of the central nervous system. Cyclodines, such as the GABA antagonists endosulphan and lindane, inhibit the calcium ion influx and Ca- and Mg-ATPase causing release of neurotransmittors. Epidemiological studies have exposed the etiological relationship between Parkinson's disease and organochlorine pollutants.

Effect in Humans

Examination of effects of different classes of pesticides leads to the conclusion that many of them are responsible for hypertension, cardiovascular disorders and other health related problems in humans. Organochlorines act as endocrine disrupting chemicals (EDCs) by interfering with molecular circuitry and function of the endocrine system. Farm workers, their families and those who pass through a region applied with pesticides can absorb a measurable quantity of pesticides. The presence of pesticide residues has been detected in blood plasma of workers in agricultural farms. Direct or indirect exposure to pesticides leads to neuromuscular disorders and stimulation of drug and steroid metabolism.

Another mode of exposure to these pesticides is through diet. Among food items, fatty food such as meat, fish, poultry, and dairy products serve as main causes. Many of the organochlorine molecules are carcinogens and neurotoxic. The hazardous nature of organochlorines was explained by citing different examples. The menace caused by endosulfan is of great concern. Endosulfan remains in the environment for longer periods and bio-accumulates in plants and animals which leads to contamination of food consumed by humans. It affects mainly the central nervous system and was found to have higher acute inhalation toxicity than dermal toxicity. Gastrointestinal absorption of endosulfan is very high.

Disproportion of thyroid hormones can lead to a variety of disorders. Serum concentrations of p-p'-DDE and HCB were found to be associated with abnormal thyroid hormone levels. p,p'-DDE was reported to increase free thyroxine (T4) and total triiodothyronine (T3) levels, and to be inversely associated with thyroid-stimulating hormone (TSH). On exposure to dioxinlike organochlorines, a dose-dependent decrease in total T4 was also reported. Organochlorine pesticides were reported to increase the risk of hormone-related cancers including breast, prostate, stomach and lung cancer. Recently dioxins have been found in human ovarian follicular fluid, which may lead to the development of endometriosis. Exposure to dioxins can cause several autoimmune diseases, including multiple sclerosis and eczema. Organochlorines can function as xenoestrogens and compounds such as TCDD, methoxychlor and alachlor were reported to exert effects on human and experimental animals due to inhibited synthesis and increased degradation of thyroid hormones.

Analysis of the National Health and Nutrition Examination Survey 1999–2004 studying the relation between organochlorine pesticides and prostate and breast cancers has shown that serum concentrations of b-HCH, trans-nonachlor, and dieldrin were significantly associated with prostate cancer prevalence. In children, exposure to dioxins showed significant positive associations with learning disability (LD). Risk of attention deficit hyperactivity disorder (ADHD) at higher levels of p,p'-DDE and PCBs exposure was reported. Prenatal exposure to p,p'-DDE and its presence in cord serum was found to lead to disappearance of neuronal development after 12 months of infant age. Epidemiological studies have shown that exposure to persistent organic pollutants, mainly organochlorine pesticides, is strongly associated with type 2 diabetes. Some persistent organic pollutants, as highly chlorinated PCBs and trans-nonachlor, were associated with the incidence of type 2 diabetes in obese people.

Selected persistent organic pollutants are reported to induce divergent actions on blood pressure, suggesting a chemical structure based association of pesticides. In a population based study, different persistent organic pollutants and pesticides were reported to be associated with liver dysfunction biomarkers such as bilirubin, ALT and ALP, suggesting that these environmental pollutants can cause adverse effects on liver functions. A study conducted in Costa Rica reported that occupational pesticide exposure to dialdrin could be partly responsible for the increased risk of Parkinson's disease seen in the population. Studies showed that the change of lipids over time, especially LDL-cholesterol, is linked to POP exposure. Increased oxidative stress markers in plasma were found to be associated with exposure of POPs and could be a causative agent for oxidative stress. Persistent organic pollutants were reported to influence the complement system, leading to activation of the immune system in humans. Detection of organochlorine pesticides from human breast milk was reported from many places in the world. In Croatia, p,p'-DDE was found to be the dominant organochlorine pesticide in human breast milk. Exposure of infants to chlordanes via breast milk was reported as a potential health risk in Korea. Another study from Korea also revealed the presence of organochlorine pesticides (OCPs) chlordanes, aldrin, dichlorodiphenyltrichloroethanes (DDTs), dieldrin, heptachlors, endrins, hexachlorocyclohexanes (HCHs), hexachlorobenzene (HCB), toxaphenes and mirex, in milk. Organochlorine pesticides HCB, β-HCH, pp'DDE, pp'DDT, pp'DDT, Σ-DDT were present in breast milk of the population in Guerrero, Mexico, proportionally to exposure.

A study conducted in China showed that prenatal exposure to DDT, β-BHC, HCB and mirex caused decrease in birth weight of infants. A number of studies were published on the effect of organochlorine pesticides on induction of diabetes mellitus in humans. A recent study reported that POP exposure is a risk factor contributing to insulin resistance. Chronic exposure to chlordecone was found to cause hypertensive disorders in pregnancy and gestational diabetes mellitus among French Caribbean women. In a study conducted in Slovakia, highly increased blood levels of diabetes (fasting glucose and insulin) and obesity markers (BMI, triglyceride and cholesterol) were found in large groups of males and females in highly polluted areas. A significant decrease in testosterone level was also observed in males. Prevalence of type 2 diabetes and exposure to persistent organic pollutants has been established. Recent studies on organochlorine pesticides have shown that β-HCH, HCB and DDT residues bio-accumulate in maternal and cord sera and from maternal blood they can be transferred through the placenta and affect thyroid hormone levels in the newborn. OC pesticides have been suggested to affect the thyroid system through gender-specific mechanisms; the extent of the effect may differ among compounds. A report from Brazil had shown that OC compounds are reported to trigger anti-androgenic effects in men and estrogenic effects in women. OC pesticide heptachlor was reported to induce mitochondria-mediated cell death via impairing electron transport chain complex III, thus acting as a neurotoxicant with possible association with Parkinson's disease.

Exposure to organochlorine pesticide residues was reported as a potential risk factor for gallstone disease in humans. Potential neurotoxic effects of organochlorine compounds were reported on early psychomotor development even at low doses. A positive correlation was observed of exposure to some OC pesticides and vitamin D deficiency in humans. Early exposure to certain environmental chemicals, especially organochlorine compounds, with endocrine-disruption activity were reported to interfere with neonatal thyroid hormone status.

Toxic Effect of Pesticides in Fauna

Wild birds are of great importance to the ecosystem. Decline in the bird community serves as an indicator of environmental pollution. Continuous use of pesticides is one of the major causes for the reduction of birds. In many cases the impact is not direct, however repetitive use of pesticides like DDT in soil is taken up by earthworms which are then ingested by birds and thus their accumulation may result in a large loss in bird population. Subsequent research has also identified other pesticides and industrial chemicals that cause mortality and reproductive impairment, which affects both embryos and adult birds. The effects on embryos include mortality or reduced hatchability, wasting syndrome and teratological effects that produce skeletal abnormalities and impaired differentiation of the reproductive and nervous systems through mechanisms of hormonal mimicking of estrogens. The range of chemical effects on adult birds covers acute mortality, sub-lethal stress, reduced fertility, suppression of egg formation, eggshell thinning and impaired incubation and chick rearing behaviors. Pesticides cause extinction, behavioral changes, loss of safe habitat and population decline in several birds. Prolonged use of pesticides causes a drastic decrease in birds like the peregrine falcon, sparrow hawk and bald eagle. The levels of organochlorines in seabird eggs were indicated by forming a deposit of pollutants in the body, thus serving as a useful indicator of environmental contamination.

Toxic Effect in Farm Animals

The prolonged use of pesticides in agriculture has caused serious health problems as these pesticides accumulate and affect the food chain. Organochlorine compounds are highly lipophilic and can accumulate in fat-rich food such as meat and milk. Pesticides are introduced into cattle mainly through fodder or contaminated water used for household and public purposes. Amphibians and insectivorous reptiles, like lizards, have an important function in linking invertebrates with vertebrates in the food chain. They serve as a food source for some organisms and are also a means by which chemical residues, especially residues of organochlorine pesticides taken in with contaminated prey, can enter food chains. Amphibians consume these pesticides by a number of ways, including inhalation, contact and through ingestion. Amphibians in open water bodies may also be exposed to pesticides due to run-off from adjacent agricultural land on which chemicals are used to control crop pests. Continuous exposure of honey bees to pesticides affects the quality of honey. The routes of honey contamination with pesticides are direct and indirect. The direct is treatment of beehives with pesticides. Wild animals, including the grasscutter (Thryonomys swinderianus), which are a good source of protein, are seriously affected by the use of pesticides. Grasscutters are a source of food for the people of Ghana in Africa. As pesticides have high effect on the animal and bird community, ultimately humans also take up pesticides as meat, milk and crops derived from these animals and plants are consumed by humans.

ALDRIN

Aldrin is an organochlorine insecticide that was widely used until the 1990s, when it was banned in most countries. Aldrin is a member of the so-called "classic organochlorines" (COC) group of pesticides. COCs enjoyed a very sharp rise in popularity during and after The Second World War. Other noteworthy examples of COCs include DDT. After research showed that organochlorines can be highly toxic to the ecosystem through bioaccumulation, most were banned from use. It is a colourless solid. Before the ban, it was heavily used as a pesticide to treat seed and soil. Aldrin and related "cyclodiene" pesticides (a term for pesticides derived from Hexachlorocyclopentadiene) became notorious as persistent organic pollutants.

Structure and Reactivity

The structure formula of aldrin is $C_{12}H_8Cl_6$. The molecule has a molecular weight of 364.896 g/mol. The melting point of aldrin is a temperature of 105 °C and the octanol-water partition coefficient is 6.5 (logP).

Pure aldrin takes form as a white crystalline powder. Though it is not soluble in water (0.003%% solubility), aldrin dissolves very well in organic solvents, such as ketones and parrafins. Aldrin decays very slowly once released into the environment. Though it is rapidly converted to dieldrin by plants and bacteria, dieldrin maintains the same toxic effects and slow decay of aldrin. Aldrin is easily transported through the air by dust particles. Aldrin does not react with mild acids or bases and is stable in an environment with a pH between 4 and 8. It is highly flammable when exposed to temperatures above 200 °C In the presence of oxidizing agents aldrin reacts with concentrated acids and phenols.

Synthesis

Aldrin is not formed in nature. It is synthesized by combining hexachlorocyclopentadiene with norbornadiene in a Diels-Alder reaction to give the adduct. In 1967, the composition of technical-grade aldrin was reported to consist of 90.5% of hexachlorohexahydrodimethanonaphthalene (HHDN).

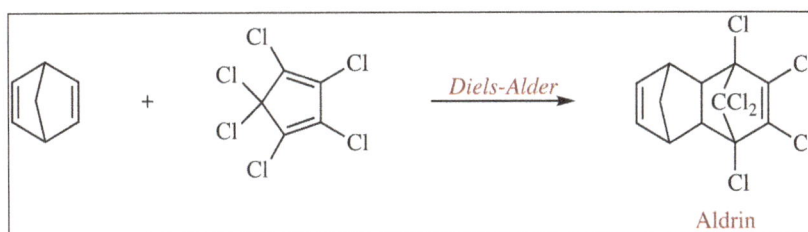

Synthesis of aldrin via a Diels-Alder reaction.

Similarly, an isomer of aldrin, known as isodrin, is produced by reaction of hexachloronobornadiene with cyclopentadiene. Isodrin is also produced as a byproduct of aldrin synthesis, with technical-grade aldrin containing about 3,5% isodrin.

Aldrin is named after the German chemist Kurt Alder, one of the coinventors of this kind of reaction. An estimated 270 million kilograms of aldrin and related cyclodiene pesticides were produced between 1946 and 1976.

Available Forms

There are multiple available forms of aldrin. One of these is the isomer isodrin, which cannot be found in nature, but needs to be synthesized like aldrin. When aldrin enters the human body or the environment it is rapidly converted to dieldrin. Degradation by ultraviolet radiation or microbes can convert dieldrin to photodieldrin and aldrin to photoaldrin.

Mechanism of Action

Even though many toxic effects of aldrin have been discovered, the exact mechanisms underlying the toxicity are yet to be determined. The only toxic aldrin induced process that is largely understood is that of neurotoxicity.

Neurotoxicity

One of the effects that intoxication with aldrin gives rise to is neurotoxicity. Studies have shown that aldrin stimulates the central nervous system (CNS), which may cause hyperexcitation and seizures. This phenomenon exerts its effect through two different mechanisms.

One of the mechanisms uses the ability of aldrin to inhibit brain calcium ATPases. These ion pumps release the nerve terminal from calcium by actively pumping it out. However, when aldrin inhibits these pumps, the intracellular calcium levels rise. This results in an enhanced neurotransmitter release.

The second mechanism makes use of aldrin's ability to block gamma-aminobutyric acid (GABA) activity. GABA is a major inhibitory neurotransmitter in the central nervous system. Aldrin induces neurotoxic effects by blocking the $GABA_A$ receptor-chloride channel complex. By blocking this receptor, chloride is unable to move into the synapse, which prevents hyperpolarization of neuronal synapses. Therefore, the synapses are more likely to generate action potentials.

Metabolism

The metabolism of oral aldrin exposure has not been studied in humans. However, animal studies are able to provide an extensive overview of the metabolism of aldrin. This data can be related to humans.

Biotransformation of aldrin starts with epoxidation of aldrin by mixed-function oxidases (CYP-450), which forms dieldrin. This conversion happens mainly in the liver. Tissues with low CYP-450 expression use the prostaglandin endoperoxide synthase (PES) instead. This oxidative pathway bisdioxygenises the arachidonic acid to prostaglandin G_2 (PGG_2). Subsequently, PGG_2 is reduced to prostaglandin H_2 (PGH_2) by hydroperoxidase.

Dieldrin can then be directly oxidized by cytochrome oxidases, which forms 9-hydroxydieldrin. An alternative for oxidation involves the opening of the epoxied ring by epoxied hydrases, which forms the product 6,7-trans-dihydroxydihydroaldrin. Both products can be conjugated to form 6,7-trans-dihydroxydihydroaldrin glucuronide and 9-hydroxydieldrin glucuronide, respectively. 6,7-trans-dihydroxydihydroaldrin can also be oxidized to form aldrin dicarboxylic acid.

Efficacy and Side Effects

Considering the toxicokinetics of aldrin in the environment, the efficacy of the compound has been determined. In addition, the adverse effects after exposure to the aldrin are demonstrated, indicating the risk regarding the compound.

Efficacy

The ability of aldrin, in its use for the control of termites, is examined in order to determine the maximum response when applied. In 1953 US researchers tested aldrin and dieldrin on terrains with rats known to carry chiggers, at a rate of 2.25 pound per acre. The Aldrin and Dieldrin treatment demonstrated a decrease of 75 times less chiggers on rats for Dieldrin treated terrains and 25 times less chiggers on the rats when treated with Aldrin. The Aldrin treatment indicate a high productivity, especially in comparison to other insecticide that were used, like DDT, Sulfur or BHC.

Adverse Effects

Exposure of Aldrin to the environment leads to the localization of the chemical compound in the air, soil, and water. Aldrin gets changed quickly to dieldrin and that compound degrades slowly, which accounts for aldrin concentrations in the environment around the primary exposure and in the plants. These concentrations can also be found in animals, which eat contaminated plants or animals that reside in the contaminated water. This biomagnification can lead to a high concentrations in their fat.

There are some reported cases of workers who developed anemia after multiple dieldrin exposures. However the main adverse effect of Aldrin and Dieldrin is in relationship to the central nervous system. The accumulated levels of Dieldrin in the body were belied to lead to convulsions. Besides that other symptoms were also reported like headaches, nausea and vomiting, anorexia, muscle twitching and mycloning jerking and EEG distortions. In all these cases removal of the source of exposure to aldrin/dieldrin led to a rapid recovery.

Toxicity

The toxicity of aldrin and dieldrin is determined by the results of several animal studies. Reports of a significant increase in workers death in relation to aldrin has not been found, although death by anemia is reported in some cases after multiple exposure to Aldrin. Immunological tests linked an antigenic response to erythrocytes coated with dieldrin in those cases. Direct dose-response relations being a cause for death are yet to be examined.

The NOAEL that was derived from rat studies:

- The minimal risk level at acute oral exposure to Aldrin is 0.002 mg/kg/day.

- The minimal risk level at intermediate exposure to Dieldrin is 0.0001 mg/kg/day.

- The minimal risk level at chronic exposure to Aldrin is 0.00003 mg/kg/day.

- The minimal risk level at chronic exposure to Dieldrin is 0.00005 mg/kg/day.

In addition to these studies, breast cancer risk studies were performed demonstrating an significant increased breast cancer risk. After comparing blood concentrations to number of lymph nodes and tumor size a 5-fold higher risk of death was determined, comparing the highest quartile range in the research to the lower quartile range. Young children are also more susceptible to the drug, causing severe generalized convulsions.

Effects on Animals

Most of the animal studies done with aldrin and dieldrin used rats. High doses of aldrin and dieldrin demonstrated neurotoxicity, but in multiple rat studies also showed an unique sensitivity of the mouse liver to dieldrin induced hepatocarcinogenicity. Furthermore Aldrin treated rats demonstrated an increased post-natal mortality, in which adults showed an increased susceptibility to the compounds compared to children in rats.

Environmental Impact and Regulation

Like related polychlorinated pesticides, aldrin is highly lipophilic. Its solubility in water is only 0.027 mg/L, which exacerbates its persistence in the environment. It was banned by the Stockholm Convention on Persistent Organic Pollutants. In the U.S., aldrin was cancelled in 1974. The substance is banned from use for plant protection by the EU.

Safety and Environmental Aspects

Aldrin has rat LD of 39 to 60 mg/kg (oral in rats). For fish however, it is extremely toxic, with an LC50 of 0.006 – 0.01 for trout and bluegill.

In the US, aldrin is considered a potential occupational carcinogen by the Occupational Safety and Health Administration and the National Institute for Occupational Safety and Health; these agencies have set an occupational exposure limit for dermal exposures at 0.25 mg/m^3 over an eighthour time-weighted average. Further, an IDLH limit has been set at 25 mg/m^3, based on acute toxicity data in humans to which subjects reacted with convulsions within 20 minutes of exposure.

It is classified as an extremely hazardous substance in the United States as defined in Section 302 of the U.S. Emergency Planning and Community Right-to-Know Act, and is subject to strict reporting requirements by facilities which produce, store, or use it in significant quantities.

DIELDRIN

Dieldrin is an organochloride originally produced in 1948 by J. Hyman & Co, Denver, as an insecticide. Dieldrin is closely related to aldrin, which reacts further to form dieldrin. Aldrin is not toxic to insects; it is oxidized in the insect to form dieldrin which is the active compound. Both dieldrin and aldrin are named after the Diels-Alder reaction which is used to form aldrin from a mixture of norbornadiene and hexachlorocyclopentadiene.

Originally developed in the 1940s as an alternative to DDT, dieldrin proved to be a highly effective insecticide and was very widely used during the 1950s to early 1970s. Endrin is a stereoisomer of dieldrin.

However, it is an extremely persistent organic pollutant; it does not easily break down. Furthermore, it tends to biomagnify as it is passed along the food chain. Long-term exposure has proven toxic to a very wide range of animals including humans, far greater than to the original insect targets. For this reason, it is now banned in most of the world.

It has been linked to health problems such as Parkinson's, breast cancer, and immune, reproductive, and nervous system damage. It is also an endocrine disruptor, acting as an estrogen and antiandrogen, and can adversely affect testicular descent in the fetus if a pregnant woman is exposed to it.

Synthesis

Dieldrin can be formed from the Diels-Alder reaction of hexachloro-1,3-cyclopentadiene with norbornadiene followed by epoxidation of the addition product.

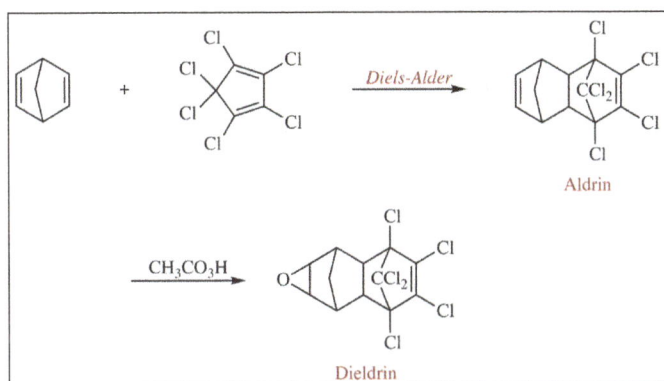

Synthesis of dieldrin.

The chemicals dieldrin and aldrin were widely applied in agricultural areas throughout the world. Both are toxic and bioaccumulative. Aldrin does break down to dieldrin in living systems, but dieldrin is known to resist bacterial and chemical breakdown processes in the environment.

Aldrin was used to control soil pests (namely termites) on corn and potato crops. Dieldrin was an insecticide used on fruit, soil, and seed. It persists in the soil with a half-life of five years at temperate latitudes. Both aldrin and dieldrin may be volatilized from sediment and redistributed by air currents, contaminating areas far from their sources. They have been measured in Arctic wildlife, suggesting long range transport from southern agricultural regions.

Both aldrin and dieldrin have been banned in most developed countries, but aldrin is still used as a termiticide in Malaysia, Thailand, Venezuela and parts of Africa. In Canada, their sale was restricted in the mid-1970s, with the last registered use of the compounds in Canada being withdrawn in 1984.

IPCS quotes the World Health Organization as stating dieldrin is prohibited for use in agriculture in, among others, Brazil, Ecuador, Finland, the German Democratic Republic, Singapore, Sweden, Yugoslavia, and the USSR. The European Community legislation prohibits the marketing of phytopharmaceutical products containing dieldrin. In Argentina, Canada, Chile, the Federal Republic of Germany, Hungary, and the USA, its use is prohibited, with some exceptions. The use of dieldrin is restricted in India, Mauritius, Togo, and the United Kingdom. Its use in industry is prohibited in Switzerland and its manufacture and use in Japan is under government control. In Finland, the only accepted use for dieldrin is as a termiticide in one glue mixture for exported plywood. India requires registration and licences for all importation, manufacture, sale, or storage.

Organochlorines and other chemicals were originally developed in the 1930s for use as insecticides and pesticides. DDT became famous worldwide in 1939 after its use in overcoming a typhus infestation in Naples. The use of organochlorines increased during the 1950s and peaked in the 1970s. Their use in Australia was dramatically lowered between the mid 1970s and the early 1980s. The first restrictions on the use of dieldrin and related chemicals in Australia were introduced in 1961-2, with registration required for their use on produce animals, such as cattle and chickens. This coincided with increasing concerns worldwide about the long-term effects of persistent pesticides. The publication of *Silent Spring* (an account of the environmental and health effects of pesticides) by Rachel Carson in 1962 was a key driving force in raising this concern. The phase-out process was driven by government bans and deregistration, in turn promoted by changing public perceptions that food containing residues of these chemicals was less acceptable and possibly hazardous to health.

Throughout this time, continuous pressure was maintained by relevant committees, for example the Technical Committee on Agricultural Chemicals (TCAC), to reduce approved organochlorine use. By 1981, the use of dieldrin worldwide was limited to sugarcane and bananas, and these uses were deregistered by 1985. In 1987, a nationwide recall system was put into place, and in December of that year, the government prohibited all imports of these chemicals into Australia without express ministerial approval. In 1994, the National Registration Authority for Agricultural and Veterinary Chemicals published a use of organochlorines in termite control, recommending the phase-out of organochlorines used in termite control upon development of viable alternatives. The same year, the Agriculture and Resource Management Council of Australia and New Zealand decided to phase out remaining organochlorine uses by 30 June 1995, with the exception of the Northern Territory. In November 1997, the use of all organochlorines other than mirex was phased out in Australia. Remaining stocks of mirex are to be used only for contained baits for termites in plantations of young trees in the Northern Territory until stocks run out, which is expected in the near future.

The recognition of negative impacts on health has stimulated the implementation of multiple legislative policies in regards to the use and disposal of organochlorine pesticides. For example, the Environment Protection (Marine) Policy 1994 became operational in May 1995 in South Australia. It dictated the acceptable concentration of toxicants such as dieldrin in marine waters and the manner in which these levels must be tested and tried.

Momentum against organochlorine and similar molecules continued to grow internationally, leading, to negotiations which matured as the Stockholm Convention on the use of persistent organic pollutants (POPs). POPs are defined as hazardous and environmentally persistent substances which can be transported between countries by the earth's oceans and atmosphere.

Most POPs (including dieldrin) bioaccumulate in the fatty tissues of humans and other animals. The Stockholm Convention banned 12 POPs, nicknamed "the dirty dozen". These include: aldicarb, toxaphene, chlordane and heptachlor, chlordimeform, chlorobenzilate, DBCP, DDT, "drins" (aldrin, dieldrin and endrin), EDB, HCH and lindane, paraquat, parathion and methyl parathion, pentachlorophenol, and 2,4,5-T. This took force on 17 May 2004. Australia ratified the Convention only three days later and became a party to it in August that year.

Well before this, Australia had been well advanced in meeting the measures agreed upon under the Convention. Production, import and use of aldrin, chlordane, DDT, dieldrin, hexachlorobenzene (HCB), heptachlor, endrin, and toxaphene are not permitted in Australia. Production and importation of polychlorinated biphenyls (PCBs) are not permitted in Australia, with the phase-out of existing PCBs being managed under the National Strategy for the Management of Scheduled Waste. This strategy also addresses how Australia will manage HCB waste and organochlorine pesticides.

Legislation in Australia on the import, use and disposal of dieldrin and other organochlorines has been extensive and covers mainly environmental and potential health impacts on the population.

ENDOSULFAN

Endosulfan is an off-patent organochlorine insecticide and acaricide that is being phased out globally. The two isomers, endo and exo, are known popularly as I and II. Endosulfan sulfate is a product of oxidation containing one extra O atom attached to the S atom. Endosulfan became a highly controversial agrichemical due to its acute toxicity, potential for bioaccumulation, and role as an endocrine disruptor. Because of its threats to human health and the environment, a global ban on the manufacture and use of endosulfan was negotiated under the Stockholm Convention in April 2011. The ban has taken effect in mid-2012, with certain uses exempted for five additional years. More than 80 countries, including the European Union, Australia, New Zealand, several West African nations, the United States, Brazil, and Canada had already banned it or announced phase-outs by the time the Stockholm Convention ban was agreed upon. It is still used extensively in India, China despite laws banning it, and few other countries. It is produced by Makhteshim Agan and several manufacturers in India and China. Although, the Supreme Court had, by an order dated 13.05.2011, put a ban on the production and sale of endosulfan in India till further orders.

Uses

Endosulfan has been used in agriculture around the world to control insect pests including whiteflies, aphids, leafhoppers, Colorado potato beetles and cabbage worms. Due to its unique mode of action, it is useful in resistance management; however, as it is not specific, it can negatively impact populations of beneficial insects. It is, however, considered to be moderately toxic to honey bees, and it is less toxic to bees than organophosphate insecticides.

Production

The World Health Organization estimated worldwide annual production to be about 9,000 metric tonnes (t) in the early 1980s. From 1980 to 1989, worldwide consumption averaged 10,500 tonnes per year, and for the 1990s use increased to 12,800 tonnes per year.

Endosulfan is a derivative of hexachlorocyclopentadiene, and is chemically similar to aldrin, chlordane, and heptachlor. Specifically, it is produced by the Diels-Alder reaction of hexachlorocyclopentadiene with *cis*-butene-1,4-diol and subsequent reaction of the adduct with thionyl chloride. Technical endosulfan is a 7:3 mixture of stereoisomers, designated α and β. α- and β-Endosulfan are configurational isomers arising from the pyramidal stereochemistry of the teravalent sulfur. α-Endosulfan is the more thermodynamically stable of the two, thus β-endosulfan irreversibly converts to the α form, although the conversion is slow.

Commercialization and Regulation

- Early 1950s: Endosulfan was developed.

- 1954: Hoechst AG (now Sanofi) won USDA approval for the use of endosulfan in the United States.

- 2000: Home and garden use in the United States was terminated by agreement with the EPA.

- 2002: The U.S. Fish and Wildlife Service recommended that endosulfan registration should be cancelled, and the EPA determined that endosulfan residues on food and in water pose unacceptable risks. The agency allowed endosulfan to stay on the US market, but imposed restrictions on its agricultural uses.

- 2007: International steps were taken to restrict the use and trade of endosulfan. It is recommended for inclusion in the Rotterdam Convention on Prior Informed Consent, and the European Union proposed inclusion in the list of chemicals banned under the Stockholm Convention on Persistent Organic Pollutants. Such inclusion would ban all use and manufacture of endosulfan globally. Meanwhile, the Canadian government announced that endosulfan was under consideration for phase-out, and Bayer Crop-Science voluntarily pulled its endosulfan products from the U.S. market but continues to sell the products elsewhere.

- 2008: In February, environmental, consumer, and farm labor groups including the Natural Resources Defense Council, Organic Consumers Association, and the United Farm Workers called on the U.S. EPA to ban endosulfan. In May, coalitions of scientists, environmental groups, and arctic tribes asked the EPA to cancel endosulfan, and in July a coalition of environmental and workers groups filed a lawsuit against the EPA challenging its 2002 decision to not ban it. In October, the Review Committee of the Stockholm Convention moved endosulfan along in the procedure for listing under the treaty, while India blocked its addition to the Rotterdam Convention.

- 2009: The Stockholm Convention's Persistent Organic Pollutants Review Committee (POPRC) agreed that endosulfan is a persistent organic pollutant and that "global action is warranted", setting the stage of a global ban. New Zealand banned endosulfan.

- 2010: The POPRC nominated endosulfan to be added to the Stockholm Convention at the Conference of Parties (COP) in April 2011, which would result in a global ban. The EPA announced that the registration of endosulfan in the U.S. will be cancelled Australia banned the use of the chemical.

- 2011: The Supreme Court of India banned manufacture, sale, and use of toxic pesticide endosulfan in India. The apex court said the ban would remain effective for eight weeks during which an expert committee headed by DG, ICMR, will give an interim report to the court about the harmful effect of the widely used pesticide.

- 2011: The Argentinian Service for Sanity and Agroalimentary Quality (SENASA) decided on August 8 that the import of endosulfan into the South American country will be banned from July 1, 2012 and its commercialization and use from July 1, 2013. In the meantime, a reduced quantity can be imported and sold.

Health Effects

Endosulfan is alleged to be responsible for many fatal pesticide poisoning incidents around the world by NGOs opposing pesticide usage. Endosulfan is also a xenoestrogen—a synthetic substance that imitates or enhances the effect of estrogens—and it can act as an endocrine disruptor, causing reproductive and developmental damage in both animals and humans. It has also been found to act as an aromatase inhibitor. Whether endosulfan can cause cancer is debated. With regard to consumers' intake of endosulfan from residues on food, the Food and Agriculture Organization of United Nations has concluded that long-term exposure from food is unlikely to present a public health concern, but short-term exposure can exceed acute reference doses.

Toxicity

Endosulfan is acutely neurotoxic to both insects and mammals, including humans. The US EPA classifies it as Category I: "Highly Acutely Toxic" based on a LD_{50} value of 30 mg/kg for female rats, while the World Health Organization classifies it as Class II "Moderately Hazardous" based on a rat LD_{50} of 80 mg/kg. It is a GABA-gated chloride channel antagonist, and a Ca^{2+}, Mg^{2+} ATPase inhibitor. Both of these enzymes are involved in the transfer of nerve impulses. Symptoms of acute poisoning include hyperactivity, tremors, convulsions, lack of coordination, staggering, difficulty breathing, nausea and vomiting, diarrhea, and in severe cases, unconsciousness. Doses as low as 35 mg/kg have been documented to cause death in humans, and many cases of sublethal poisoning have resulted in permanent brain damage. Farm workers with chronic endosulfan exposure are at risk of rashes and skin irritation.

EPA's acute reference dose for dietary exposure to endosulfan is 0.015 mg/kg for adults and 0.0015 mg/kg for children. For chronic dietary expsoure, the EPA references doses are 0.006 mg/(kg·day) and 0.0006 mg/(kg·day) for adults and children, respectively.

Endocrine Disruption

Theo Colborn, an expert on endocrine disruption, lists endosulfan as a known endocrine disruptor, and both the EPA and the Agency for Toxic Substances and Disease Registry consider endosulfan to be a potential endocrine disruptor. Numerous *in vitro* studies have documented its potential to disrupt hormones and animal studies have demonstrated its reproductive and developmental toxicity, especially among males. A number of studies have documented that it acts as an antiandrogen in animals. Endosulfan has shown to affect crustacean molt cycles, which are important biological and endocrine-controlled physiological processes essential for the crustacean growth and reproduction. Environmentally relevant doses of endosulfan equal to the EPA's safe dose of 0.006 mg/kg/day have been found to affect gene expression in female rats similarly to the effects of estrogen. It is not known whether endosulfan is a human teratogen (an agent that causes birth defects), though it has significant teratogenic effects in laboratory rats. A 2009 assessment concluded the endocrine disruption in rats occurs only at endosulfan doses that cause neurotoxicity.

Reproductive and Developmental Effects

Some studies have documented that endosulfan can also affect human development. Researchers studying children from many villages in Kasargod District, Kerala, India, have linked endosulfan exposure to delays in sexual maturity among boys. Endosulfan was the only pesticide applied to cashew plantations in the villages for 20 years, and had contaminated the village environment. The researchers compared the villagers to a control group of boys from a demographically similar village that lacked a history of endosulfan pollution. Relative to the control group, the exposed boys had high levels of endosulfan in their bodies, lower levels of testosterone, and delays in reaching sexual maturity. Birth defects of the male reproductive system, including cryptorchidism, were also more prevalent in the study group. The researchers concluded, "our study results suggest that endosulfan exposure in male children may delay sexual maturity and interfere with sex hormone synthesis." Increased incidences of cryptorchidism have been observed in other studies of endosulfan exposed populations.

A 2007 study by the California Department of Public Health found that women who lived near farm fields sprayed with endosulfan and the related organochloride pesticide dicofol during the first eight weeks of pregnancy are several times more likely to give birth to children with autism. However a 2009 assessment concluded that epidemiology and rodent studies that suggest male reproductive and autism effects are open to other interpretations, and that developmental or reproductive toxicity in rats occurs only at endosulfan doses that cause neurotoxicity.

Cancer

Endosulfan is not listed as known, probable, or possible carcinogen by the EPA, IARC, or other agencies. No epidemiological studies link exposure to endosulfan specifically to cancer in humans, but *in vitro* assays have shown that endosulfan can promote proliferation of human breast cancer cells. Evidence of carcinogenicity in animals is mixed.

In a 2016 study by the Department of Biochemistry, Indian Institute of Science, Bangalore published in *Carcinogenesis*, endosulfan was found to induce reactive oxygen species (ROS) in a concentration and time-dependent manner leading to double-stranded breaks in the DNA and also found to favour subsequent erroneous DNA repair.

Environmental Fate

Endosulfan is a ubiquitous environmental contaminant. The chemical is semivolatile and persistent to degradation processes in the environment. Endosulfan is subject to long-range atmospheric transport, *i.e.* it can travel long distances from where it is used. Thus, it occurs in many environmental compartments. For example, a 2008 report by the National Park Service found that endosulfan commonly contaminates air, water, plants, and fish of national parks in the US. Most of these parks are far from areas where endosulfan is used. Endosulfan has been found in remote locations such as the Arctic Ocean, as well as in the Antarctic atmosphere. The pesticide has also been detected in dust from the Sahara Desert collected in the Caribbean after being blown across the Atlantic Ocean. The compound has been shown to be one of the most abundant organochlorine pesticides in the global atmosphere.

The compound breaks down into endosulfan sulfate, endosulfan diol, and endosulfan furan, all of which have structures similar to the parent compound and, according to the EPA, "are also of toxicological concern. The estimated half-lives for the combined toxic residues (endosulfan plus endosulfan sulfate) range from roughly 9 months to 6 years." The EPA concluded, "based on environmental fate laboratory studies, terrestrial field dissipation studies, available models, monitoring studies, and published literature, it can be concluded that endosulfan is a very persistent chemical which may stay in the environment for lengthy periods of time, particularly in acid media." The EPA also concluded, "endosulfan has relatively high potential to bioaccumulate in fish." It is also toxic to amphibians; low levels have been found to kill tadpoles.

In 2009, the committee of scientific experts of the Stockholm Convention concluded, "endosulfan is likely, as a result of long range environmental transport, to lead to significant adverse human health and environmental effects such that global action is warranted." In May 2011, the Stockholm Convention committee approved the recommendation for elimination of production and use of endosulfan and its isomers worldwide. This is, however, subject to certain exemptions. Overall, this will lead to its elimination from the global markets.

HEPTACHLOR

Heptachlor is an organochlorine compound that was used as an insecticide. Usually sold as a white or tan powder, heptachlor is one of the cyclodiene insecticides. In 1962, Rachel Carson's Silent Spring questioned the safety of heptachlor and other chlorinated insecticides. Due to its highly stable structure, heptachlor can persist in the environment for decades. In the United States, the Environmental Protection Agency has limited the sale of heptachlor products to the specific application of fire ant control in underground transformers. The amount that can be present in different foods is regulated.

Synthesis

Analogous to the synthesis of other cyclodienes, heptachlor is produced via the Diels-Alder reaction of hexachlorocyclopentadiene and cyclopentadiene. The resulting adduct is chlorinated followed by treatment with hydrogen chloride in nitromethane in the presence of aluminum trichloride or with iodine monochloride.

Compared to chlordane, it is about 3–5 times more active as an insecticide, but more inert chemically, being resistant to water and caustic alkalies.

Metabolism

Soil microorganisms transform heptachlor by epoxidation, hydrolysis, and reduction. When the compound was incubated with a mixed culture of organisms, chlordene (hexachlorocyclopentadine, its precursor) formed, which was further metabolized to chlordene epoxide. Other metabolites include 1-hydroxychlordene, 1-hydroxy-2,3-epoxychlordene, and heptachlor epoxide. Soil microorganisms hydrolyze heptachlor to give ketochlordene. Rats metabolize heptachlor to the epoxide 1-exo-1-hydroxyheptachlor epoxide and 1,2-dihydrooxydihydrochlordene. When heptachlor epoxide was incubated with microsomal preparations form liver of pigs and from houseflies, the products found were diol and 1-hydroxy-2,3-epoxychlordene. The metabolic scheme in rats shows two pathways with the same metabolite. The first involves following scheme: heptachlor heptachlor epoxide dehydrogenated derivative of 1-exo-hydroxy-2,3-exo-epoxychlordene 1,2-dihydrooxydihy-drochlordene. The second involves: heptachlor 1-exo-hydroxychlordene 1-exo-hydroxy, 2,3-exo-epoxychlordene 1,2-dihydrooxydihydrochlordene.

Environmental Impact

Heptachlor is persistent organic pollutant (POP). It has a half life of ~1.3-4.2 days (air), ~0.03-0.11 years (water), and ~0.11-0.34 years (soil). One study described its half life to be 2 years and claimed that its residues could be found in soil 14 years after its initial application. Like other POPs, heptachlor is lipophilic and poorly soluble in water (0.056 mg/L at 25 °C), thus it tends to accumulate in the body fat of humans and animals.

Heptachlor epoxide is more likely to be found in the environment than its parent compound. The epoxide also dissolves more easily in water than its parent compound and is more persistent. Heptachlor and its epoxide absorb to soil particles and evaporate.

Toxicity of Heptachlor and Related Derivatives

The range of oral rat LD_{50} values are 40 mg/kg to 162 mg/kg. Daily oral doses of heptachlor at 50 and 100 mg/kg were found to be lethal to rats after 10 days. For heptachlor epoxide, the oral LD_{50} values ranging from 46.5 to 60 mg/kg. With rat oral of LD_{50} 47mg/kg, heptachlor epoxide is more toxic. A product of hydrogenation of heptachlor, β-dihydroheptachlor, has high insecticidal activity and low mammalian toxicity, rat oral LD_{50}>5,000mg/kg.

Human Impact

Humans may be exposed to heptachlor through drinking water and foods, including breast milk. Heptachlor epoxide is derived from a pesticide that was banned in the U.S. in the 1980s. It is still found in soil and water supplies and can turn up in food. It can be passed along in breast milk.

The International Agency for Research on Cancer and the EPA have classified the compound as a possible human carcinogen. Animals exposed to heptachlor epoxide during gestation and infancy are found to have changes in nervous system and immune function. Exposure to higher doses of heptachlor in newborn animals leads to decreased body weight and death.

The U.S. EPA MCL for drinking water is 0.0004 mg/L for heptachlor and 0.0002 mg/L for heptachlor epoxide. The U.S. FDA limit on food crops is 0.01 ppm, in milk 0.1 ppm, and on edible seafoods 0.3 ppm. The Occupational Safety and Health Administration has limit of 0.5 mg/m³ (cubic meter of workplace air) for 8-hour shifts and 40-hour work weeks.

An ATSDR report in 1993 found no studies with respect to death in humans after oral exposure to heptachlor or heptachlor epoxide.

Chemical Properties

The octanol-water partition coefficient (K_{ow}) of heptachlor is ~$10^{5.27}$. Henry's Law constant is $2.3 \cdot 10^{-3}$ atm-m³/mol and the vapor pressure is $3 \cdot 10^{-4}$ mmHg at 20 °C.

METHOXYCHLOR

Methoxychlor is a contact and stomach insecticide effective against a wide range of pests encountered in agriculture, households, and ornamental plantings. It is registered for use on fruits, vegetables, forage crops and on shade trees. Methoxychlor is also registered for veterinary use as a poison to kill parasites on dairy and beef cattle.

Methoxychlor is one of a few organochlorine pesticides that have seen an increase in use since the ban on DDT in 1972. This is due to its relatively low toxicity and relatively short persistence in biological systems. Methoxychlor is generally used as pesticide.

Toxicological Effects

Acute Toxicity

Methoxychlor is classified as slightly toxic and carries the signal word caution. It has a very low toxicity. The oral LD50 for rats is 5,000 to 6,000 mg/kg and 2,000 mg/kg for mice. A 50% mortality was not achieved for monkeys at 2,500 mg/kg or for hamsters at 2000 mg/kg. The lowest oral dose that can cause lethal effects for humans is estimated to be 6,400 mg/kg and the lowest dose through the skin that produces toxic effects in humans is 2,400 mg/kg based on behavioral symptoms. Rabbit skin dosed at 2,800 mg/kg produced no symptoms. Symptoms close to the lethal dose include Central Nervous System depression, progressive weakness and diarrhea. Extremely high doses can cause death within 36 to 48 hours.

Chronic Toxicity

Rats fed from very low to high doses of methoxychlor (10 to 2,000 mg/kg) for two years had growth retardation above 200 ppm but no tissue damage from the methoxychlor. Human volunteers taking oral doses of 0.5 to 2.0 mg/kg/day for six weeks had no adverse effects measured by routine enzyme (biochemical) or (blood) hematologic parameters. Loss of body weight and growth retardation were the most frequent effects in lab animal studies. These effects were attributed to food refusal rather than to methoxychlor toxicity.

Reproductive Effects

Rats fed low doses (about 50 mg/kg/day) in their diet had normal reproduction but slightly higher doses (150 mg/kg/day) fewer animals mated and many did not produce litters. At about 250 mg/kg/day none of the rats had litters or embryo implantation. In another study male rats given 100 to 200 mg/kg/day suffered arrested sperm production after 70 consecutive days and females rats produced ovarian effects. Chronic exposure to this pesticide may present a reproductive risk to humans.

Teratogenic Effects

When a methoxychlor formulation containing 50% active ingredient and 50% unknown compounds was administered to pregnant female rats, adverse effects in the fetus occured only at doses large enough to be toxic to the dams. These effects were thought to be due to the disruption of the maturation process rather than due to the direct toxic effects of methoxychlor. At 400 mg/kg, the pesticide killed rat embryos and at 200 mg/kg there was increased incidence of resorption, small litter size, and low fetal weights. This suggests that there may be a potential risk to human development following chronic exposure.

Mutagenic Effects

Most mutation assays have proven to be negative. There is no convincing evidence that methoxychlor is toxic to genetic material.

Carcinogenic Effects

Two strains of mice were fed diets containing low levels (40 mg/kg) methoxychlor for two years. There was no significant incidence of liver tumors but one strain did have increased testicular tumors. After evaluating the data, National Cancer Institute and the International Agency for Research on Cancer both conclude that methoxychlor is not an animal carcinogen. The U.S. EPA has not made an official determination on the carcinogenic status of the compound.

Organ Toxicity

Chronic effects include liver cell degeneration and kidney damage. Death is usually due to respiratory failure from paralysis in the brain. Central nervous depression is more prominent than excitation.

Methoxychlor does not accumulate to any significant degree in fat or other tissues of mammals. At high dietary doses, low levels of methoxychlor were detected in the fat of rats though it cleared the body readily after dietary intake stopped (two weeks). Mice excreted 98.3% of a 50 mg/kg dose in the urine and feces within 24 hours. When rats were injected with 3 mg/kg, 50% was excreted in the feces and 5 to 10% in the urine in four days. The major metabolites in mouse feces and urine were the monophenol and bisphenol. Other metabolites were present also but methoxychlor itself does not appear to undergo dehydrochlorination.

Lactating cows treated twice in 14 days with sprays of 0.25 to 0.5% (2 quarts per animal) had residues of 2 to 3 ppm in milk. After 14 days, levels were at the limit of detection (0.005 ppm).

Ecological Effects

Methoxychlor shows low toxicity to mallards, Japanese quail, pheasants and bobwhite quail. No mortality occurred in these species after being exposed at 250 mg/kg in their diets for five days. Most fish, however, are sensitive to the pesticide. The 96-hour LC50 for fish ranges from 1.7 ppb for Atlantic salmon to 5,200 ppm for channel catfish. The bioconcentration factors for fish ranges from 138 in sheepshead minnows to 8,300 in the fathead minnow. Bioconcentration factors were the highest in the mussel (12,000) and in the snail (8,570). This indicates that methoxychlor would accumulate in aquatic organisms that do not rapidly metabolize the compound. Fish do metabolize methoxychlor fairly rapidly and thus tend not to accumulate it appreciably.

Environmental Fate

Methoxychlor is very persistent in soil and its half-life is greater than six months. However, rates may be as fast as one week in some instances. The chemical is tightly bound to soil and is insoluble in water, so it leaches slowly, if at all. Methoxychlor degrades much more rapidly in soil that has a supply of oxygen (aerobic) than in soil without oxygen (anaerobic). Any movement of the pesticide is expected to take place while attached to suspended soil particles in runoff. In the EPA pilot groundwater survey, methoxychlor was found in a number of wells in New Jersey (not quantified) and at extremely low concentrations (from 0.1 to 1.0 ppt) in water from the Niagara River, the James river, and a Lake Michigan tributary. Many other rivers tested throughout the United States did not contain methoxychlor.

In water the major products of breakdown in a neutral solution are anisoin, anisil, and DMDE. The half-life in distilled water is 37 days but in some river waters the half-life is as rapid as two to five hours. Methoxychlor evaporates very slowly, but the evaporation may contribute to the cycling of the product in the environment.

On mature soybean foliage, the washoff rate was 8% per cm of rainfall with a total of 33.5% washoff for a season. Dislodgeable residues account for less than 1% of the amount applied.

Table: Physical properties.

Chemical name	1,1'-(2,2,2-trichloro-ethylidene)bis[4-methoxybenzene]
Chemical class/use	diphenyl alkane insecticide
Solubility in water	0.1 mg/l
Solubility in other solvents	chloroform 44 g/100 g; methanol 5 g/100 g
Melting Point	86-88 degrees C
Vapor Pressure	very low
Partition Coefficient	3.05 to 4.30 (octanol/water)

MIREX

Mirex is an organochloride that was commercialized as an insecticide and later banned because of its impact on the environment. This white crystalline odorless solid is a derivative of cyclopentadiene. It was popularized to control fire ants but by virtue of its chemical robustness and lipophilicity it was recognized as a bioaccumulative pollutant. The spread of the red imported fire ant was encouraged by the use of Mirex, which also kills native ants that are highly competitive with the fire ants. The United States Environmental Protection Agency prohibited its use in 1976. It is prohibited by the Stockholm Convention on Persistent Organic Pollutants.

Production and Applications

Mirex was first synthesized in 1946, but was not used in pesticide formulations until 1955. Mirex was produced by the dimerization of hexachlorocyclopentadiene in the presence of aluminium chloride.

Mirex is a stomach insecticide, meaning that it must be ingested by the organism in order to poison it. The insecticidal use was focused on Southeastern United States to control the imported fire ants *Solenopsis saevissima richteri* and *Solenopsis invicta*. Approximately 250,000 kg of mirex was applied to fields between 1962-75 (US NRC, 1978). Most of the mirex was in the form of "4X mirex bait," which consists of 0.3% mirex in 14.7% soybean oil mixed with 85% corncob grits. Application of the 4X bait was designed to give a coverage of 4.2 g mirex/ha and was delivered by aircraft, helicopter or tractor. 1x and 2x bait were also used. Use of mirex as a pesticide was banned in 1978. The Stockholm Convention banned production and use of several persistent organic pollutant, and Mirex is one of the "dirty dozen".

Degradation

Characteristic of chlorocarbons, mirex does not burn easily; combustion products are expected to include carbon dioxide, carbon monoxide, hydrogen chloride, chlorine, phosgene, and other organochlorine species. Slow oxidation produces chlordecone ("Kepone"), a related insecticide that is also banned in most of the western world, but more readily degraded. Sunlight degrades mirex primarily to photomirex (8-monohydromirex) and later partly to 2,8-dihydromirex.

Photomirex.   2,8-dihydromirex.

Mirex is highly resistant to microbiological degradation. It only slowly dechlorinates to a monohydro derivative by anaerobic microbial action in sewage sludge and by enteric bacteria.

Bioaccumulation and Biomagnification

Mirex is highly cumulative and amount depends upon the concentration and duration of exposure. There is evidence of accumulation of mirex in aquatic and terrestrial food chains to harmful levels. After 6 applications of mirex bait at 1.4 kg/ha, high mirex levels were found in some species; turtle fat contained 24.8 mg mirex/kg, kingfishers, 1.9 mg/kg, coyote fat, 6 mg/kg, opossum fat, 9.5 mg/kg, and racoon fat, 73.9 mg/kg. In a model ecosystem with a terrestrial-aquatic interface, sorgum seedlings were treated with mirex at 1.1 kg/ha. Caterpillars fed on these seedlings and their faeces contaminated the water which contained algae, snails, Daphnia, mosquito larvae, and fish. After 33 days, the ecological magnification value was 219 for fish and 1165 for snails.

Although general environmental levels are low, it is widespread in the biotic and abiotic environment. Being lipophilic, Mirex is strongly adsorbed on sediments.

Safety

Mirex is only moderately toxic in single-dose animal studies (oral LD_{50} values range from 365–3000 mg/kg body weight). It can enter the body via inhalation, ingestion, and via the skin. The most sensitive effects of repeated exposure in animals are principally associated with the liver, and these effects have been observed with doses as low as 1.0 mg/kg diet (0.05 mg/kg body weight per day), the lowest dose tested. At higher dose levels, it is fetotoxic (25 mg/kg in diet) and teratogenic (6.0 mg/kg per day). Mirex was not generally active in short-term tests for genetic activity. There is sufficient evidence of its carcinogenicity in mice and rats. Delayed onset of toxic effects and mortality is typical of mirex poisoning. Mirex is toxic for a range of aquatic organisms, with crustacea being particularly sensitive.

Mirex induces pervasive chronic physiological and biochemical disorders in various vertebrates. No acceptable daily intake (ADI) for Mirex has been advised by FAO/WHO. IARC (1979) evaluated mirex's carcinogenic hazard and concluded that "there is sufficient evidence for its carcinogenicity to mice and rats. In the absence of adequate data in humans, based on above result it can be said, that it has carcinogenic risk to humans". Data on human health effects do not exist.

Health Effects

Per a 1995 ATSDR report Mirex caused fatty changes in the livers, hyperexcitability and convulsion, and inhibition of reproduction in animals. It is a potent endocrine disruptor, interfering with estrogen-mediated functions such as ovulation, pregnancy, and endometrial growth. It also induced liver cancer by interaction with estrogen in female rodents.

References

- Centers for Disease Control and Prevention (4 April 2011). "Aldrin". NIOSH Pocket Guide to Chemical Hazards. Retrieved 13 November 2013

- PMC5464684: ncbi.nlm.nih.gov, Retrieved 5 May, 2019

- Robert L. Metcalf "Insect Control" in Ullmann's Encyclopedia of Industrial Chemistry, Wiley-VCH, Weinheim, 2002. Doi:10.1002/14356007.a14_263

- "Bayer to stop selling endosulfan". Australian Broadcasting Corporation. July 17, 2009. Retrieved 2009-07-17

- Methoxychlor-ext, haloxyfop-methylparathion, extoxnet, profiles: .cce.cornell.edu, Retrieved 6 June, 2019

- Robert L. Metcalf "Insect Control" in Ullmann's Encyclopedia of Industrial Chemistry" Wiley-VCH, Weinheim, 2002. Doi:10.1002/14356007.a14_263

5

Health and Environmental Effects

Pesticides and insecticides have a widespread impact on the environment as well as human health. They have devastating effects on soil, water, air and wildlife and cause various skin and eye diseases in humans. The chapter closely examines these environmental and health impacts of pesticides and insecticides to gain a better understanding of the subject.

ENVIRONMENTAL IMPACT OF PESTICIDES

Pesticides are sprayed over large areas of land, they have a widespread impact on the environment. Research has shown, for example, that over 95% of herbicides and over 98% of insecticides do not reach the targeted pest. This is because pesticides are applied over large tracts of land and carried away by wind and water runoff. As these chemicals travel to other areas, they affect a number of plant and animal species. Additionally, storage, transportation, and production allow some quantities of pesticides to be introduced to the environment.

While research concerning the exact impact of pesticides on the environment is varied, it has increased over the last few decades. As a result of some of this research, both the United States and the European Union have stopped using organophosphate and carbamate insecticides, some of the most toxic of all pesticides. Companies have begun developing pesticides with reduced side effects for non-target species as well.

Impact of Pesticides on Soil

Once applied to crops, pesticides work their way into the soil, where it has devastating effects. Perhaps the most detrimental of these effects is that pesticide causes biodiversity loss in soil. This means the soil has a lower quality overall and is less fertile. Additionally, it removes a large percentage of organic matter. Organic matter helps soil retain water, which can be extremely helpful to agricultural workers, particularly during droughts. This lack of organic matter also allows pesticides to continue to build up in the soil instead of breaking down the chemicals. Less fertile soils mean less plant growth, which, in turn, means farmers must use increased quantities of fertilizer for successful crop yields.

Impact of Pesticides on Water

Pesticides seep into the soil and find their way into groundwater. Additionally, they may be washed into nearby streams and rivers. In fact, research has found that every stream and around 90% of all water wells are polluted with pesticides in the US. Rain and groundwater sources have also been found to be contaminated. Several countries around the world, including the US and the UK, have passed drinking water safety laws in an attempt to regulate and reduce the amount of pesticides found in public water systems.

Impact of Pesticides on Air

Pesticides do not only collect on plants, seep into the ground, and wash away into nearby waterways. These chemicals are also easily carried on the wind to other, non-agricultural areas, in a phenomenon known as pesticide drift. Pesticide drift occurs when pesticide is sprayed on crops and carried off by the wind before reaching the plants or when it undergoes volatilization. Herbicide (or pesticide) volatilization is what happens when the chemical reaches its intended destination and later evaporates into the air, being carried downwind. It is more common in warmer climates and seasons when evaporation occurs at a faster rate, preventing the pesticide from being absorbed into the ground.

Once the chemicals leave the intended target, they can be carried across long distances, potentially entering delicate ecosystems. The distance that these pesticides can travel depends on wind speed, relative humidity levels, and external temperatures. This means that warmer summer temperatures typically result in increased pesticide concentrations in the air, which are then introduced to human and animal respiratory systems. Some pesticides also emit volatile organic compounds that react with other chemicals in the atmosphere and create tropospheric ozone, a greenhouse gas that affects how long methane and other hydrocarbons remain in the atmosphere. In order to prevent pesticides from being carried through the air, many countries have implemented regulations that require windbreaks or buffer zones around targeted crops. These can take the form of tall pine trees planted around the agricultural land or empty fields surrounding the pesticide-treated area.

Impact of Pesticides on Wildlife

Pesticide use affects both plants and animals. Chemicals reduce nitrogen fixation, the symbiotic relationship between nitrogen fixing bacteria and plants that is required for proper plant growth. A reduction in nitrogen fixation results in reduced crop yield, particularly in legume type plants. When this occurs, additional fertilizer must be applied to the fields. Pesticide use is also directly linked to the constantly declining bee population, a species which is vital to plant pollination. In fact, researchers have studied this effect, known as pollinator decline, in order to understand Colony Collapse Disorder. This disorder occurs when bee colonies are exterminated without prior indicators of population decline. The US Department of Agriculture has released an estimate suggesting that US farmers lose approximately $200 million annually due to reduced pollination.

Wildlife other than plants and bees are affected by pesticide use as well. Many animal species may inadvertently ingest pesticides after eating food that has come into contact with the chemicals. Humans also have this risk. Because of its ability to be carried over long distances, these chemicals may also reach other ecosystems and cause significant damage. Pesticides have been linked to reduced plant growth in non-target areas, which leaves wildlife with little to no food source. These animals are then forced to leave their territory in search of sustenance or die due to a lack of available food. Additionally, pesticides are carried up the food chain when animals consume pesticide-contaminated foods. This was seen in North America with birds of prey, particularly eagles. These birds were consuming pesticide-contaminated fish. The pesticide underwent bioaccumulation and was passed on in a more concentrated form to the hatchlings of these birds, causing them to die at a young age or while still incubating in the egg.

Plants, birds, fish, reptiles, amphibians, and mammals (including humans) have all been affected by pesticide use. It appears that this man-made chemical was invented with the intention of improving and increasing crop yields to ensure continued health of the human population. Unfortunately, its use has come with unintended and fatal consequences. Governments around the world need to act in order to control pesticide application and prevent some of its harmful and increasingly common side effects.

HEALTH EFFECTS OF PESTICIDES

Pesticides are toxic, they are also potentially hazardous to humans, animals, other organisms, and the environment. Therefore, people who use pesticides or regularly come in contact with them must understand the relative toxicity, potential health effects, and preventative measures to reduce exposure to the products they use.

Pesticide Toxicity and Exposure

Hazard, or risk, of using pesticides is the potential for injury, or the degree of danger involved in using a pesticide under a given set of conditions. Hazard depends on the toxicity of the pesticide and the amount of exposure to the pesticide and is often illustrated with the following equation:

$$Hazard = Toxicity \times Exposure$$

The toxicity of a pesticide is a measure of its capacity or ability to cause injury or illness. The toxicity of a particular pesticide is determined by subjecting test animals to varying dosages of the active ingredient (a.i.) and each of its formulated products. The active ingredient is the chemical component in the pesticide product that controls the pest. By understanding the difference in toxicity levels of pesticides, a user can minimize the potential hazard by selecting the pesticide with the lowest toxicity that will control the pest.

Applicators may have little or no control over the availability of low-toxicity products or the toxicity of specific formulated products. However, applicators can minimize or nearly eliminate exposure - and thus reduce hazard - by following the label instructions, using personal protective clothing and equipment (PPE), and handling the pesticide properly. For example, more than 95 percent of all pesticide exposures come from dermal exposure, primarily to the hands and forearms. By wearing a pair of unlined, chemical-resistant gloves, this type of exposure can be nearly eliminated.

Acute Toxicity and Acute Effects

Acute toxicity of a pesticide refers to the chemical's ability to cause injury to a person or animal from a single exposure, generally of short duration. The harmful effects that occur from a single exposure by any route of entry are termed "acute effects." The four routes of exposure are dermal (skin), inhalation (lungs), oral (mouth), and the eyes. Acute toxicity is determined by examining the dermal toxicity, inhalation toxicity, and oral toxicity of test animals. In addition, eye and skin irritation are also examined.

Acute toxicity is measured as the amount or concentration of a toxicant - the a.i. - required to kill 50 percent of the animals in a test population. This measure is usually expressed as the LD_{50} (lethal dose 50) or the LC_{50} (lethal concentration 50). Additionally, the LD_{50} and LC_{50} values are based on a single dosage and are recorded in milligrams of pesticide per kilogram of body weight (mg/kg) of the test animal or in parts per million (ppm). LD_{50} and LC_{50} values are useful in comparing the toxicities of different active ingredients and different formulations containing the same active ingredient. The lower the LD_{50} or LC_{50} value of a pesticide product, the greater its toxicity to humans and animals. Pesticides with a high LD_{50} are the least toxic to humans if used according to the directions on the product label.

Chronic Toxicity and Chronic Effects

The chronic toxicity of a pesticide is determined by subjecting test animals to long-term exposure to the active ingredient. Any harmful effects that occur from small doses repeated over a period of time are termed "chronic effects." Suspected chronic effects from exposure to certain pesticides include birth defects, toxicity to a fetus, production of benign or malignant tumors, genetic changes, blood disorders, nerve disorders, endocrine disruption, and reproduction effects. The chronic toxicity of a pesticide is more difficult than acute toxicity to determine through laboratory analysis.

Pesticide Signal Words

Products are categorized on the basis of their relative acute toxicity (their LD_{50} or LC_{50} values). Pesticides that are classified as highly toxic (Toxicity Category I) on the basis of either oral, dermal, or inhalation toxicity must have the signal words DANGER and POISON printed in red with a skull and crossbones symbol prominently displayed on the front panel of the package label. The Spanish equivalent for DANGER, "PELIGRO," must also appear on the labels of highly toxic chemicals. The acute (single dosage) oral LD_{50} for pesticide products in this group ranges from a trace amount to 50 mg/kg. For example, exposure of a few drops of a material taken orally could be fatal to a 150-pound person.

Some pesticide products have just the signal word DANGER, which tells you nothing about the acute toxicity, just that the product can cause severe eye damage or severe skin irritation.

Pesticide products considered moderately toxic (Toxicity Category II) must have the signal word WARNING and "AVISO" (the Spanish equivalent) displayed on the product label. In this category, the acute oral LD_{50} ranges from 50 to 500 mg/kg. A teaspoon to an ounce of this material could be fatal to a 150-pound person.

Pesticide products classified as either slightly toxic or relatively nontoxic (Toxicity Categories III and IV) are required to have the signal word CAUTION on the pesticide label. Acute oral LD_{50} values in this group are greater than 500 mg/kg. An ounce or more of this material could be fatal to a 150-pound person.

Table summarizes the LD_{50} and LC_{50} values for each route of exposure for the four toxicity categories and their associated signal word. For example, an active ingredient with a dermal LD_{50} of 1,000 mg/kg would be in Toxicity Category II with a WARNING signal word. Keep in mind that an active ingredient may have a high LD_{50} placing it in a Toxicity Category II, III, or IV but also have corrosive eye/skin effects that take priority and place it in Toxicity Category I.

Routes of Exposure	Toxicity Category			
	I	II	III	IV
Oral LD	Up to and including 50 mg/kg	50-500 mg/kg	500-5,000 mg/kg	>5,000 mg/kg
Inhalation LC	Up to and including 0.2 mg/l	0.2-2 mg/l	2-20 mg/l	>20 mg/l
Dermal LD	Up to and including 200 mg/kg	200-2,000 mg/kg	2,000-20,000 mg/kg	>20,000 mg/kg
Eye Effects	Corrosive corneal opacity not reversible within 7 days	Corneal opacity reversible within 7 days; irritation persisting for 7 days	No corneal opacity; irritation reversible within 7 days	No irritation
Skin Effects	Corrosive	Severe irritation at 72 hours	Moderate irritation at 72 hours	Mild or slight irritation at 72 hours
Signal Word	Danger Poison	Warning	Caution	Caution

All pesticide toxicity values, including the LD_{50}, can be found on the product's Material Safety Data Sheet (MSDS). Pesticide labels and MSDS can be obtained from retailers or manufactures. In addition, most products also have information that can be found on the Internet.

Symptoms of Pesticide Poisoning

The symptoms of pesticide poisoning can range from a mild skin irritation to coma or even death. Different classes or families of chemicals cause different types of symptoms. Individuals also vary in their sensitivity to different levels of these chemicals. Some people may show no reaction to an exposure that may cause severe illness in others. Because of potential health concerns, pesticide users and handlers must recognize the common signs and symptoms of pesticide poisoning.

The effects, or symptoms, of pesticide poisoning can be broadly defined as either topical or systemic. Topical effects generally develop at the site of pesticide contact and are a result of either the pesticide's irritant properties (either the active and inert ingredient) or an allergic response by the victim. Dermatitis, or inflammation of the skin, is accepted as the most commonly reported topical effect associated with pesticide exposure. Symptoms of dermatitis range from reddening of the skin to rashes and blisters.

Some individuals tend to cough, wheeze, or sneeze when exposed to pesticide sprays. Some individuals react to the strong odor and irritating effects of petroleum distillates used as carriers in pesticide products. One symptom is that the eyes, mucous membranes of the nose, and even the sensitive linings of the mouth and back of the throat feel raw and scratchy. This symptom usually subsides within a few minutes after a person is removed from the exposure to the irritant. However, a reaction to a pesticide product that causes someone not only to sneeze and cough but also to develop severe acute respiratory symptoms is more likely to be a true hypersensitivity or allergic reaction. Symptoms of a true allergic reaction range from reddening and itching of the eyes and skin to respiratory discomfort often resembling an asthmatic condition.

Systemic effects are quite different from topical effects. They often occur away from the original point of contact as a result of the pesticide being absorbed into and distributed throughout the body. Systemic effects often include nausea, vomiting, fatigue, headache, and intestinal disorders. In advanced poisoning cases, the individual may experience changes in heart rate, difficulty breathing, convulsions, and coma, which could lead to death.

Seeking Medical Attention

Be alert for the early signs and symptoms of pesticide poisoning in yourself and others. These often occur immediately after exposure, but they could be delayed for up to 24 hours. If you are having symptoms but are unsure if they are pesticide related, at least notify someone in case your symptoms become worse. But when symptoms appear after contact with pesticides, you should seek medical attention immediately.

If safe to do so, take the pesticide container to the telephone. (However, if the pesticide container is contaminated, write down the product name, active ingredient(s) and percentage, and the EPA registration number.) The product label provides medical personnel information such as active ingredients, an antidote, and an emergency contact number for the manufacturer. If the Material Safety Data Sheet is available, take this also because it contains additional information for medical personnel.

If you must go to the hospital or doctor's office, take the entire pesticide container, including the label, with you. In order to avoid inhaling fumes or spilling the contents, make sure the container is tightly sealed and place into a plastic bag if possible. The pesticide container should never be placed in the enclosed passenger section of your vehicle.

Harmful Effects of some Pesticide Families

Fungicides

The acute toxicity of fungicides to humans is generally considered to be low, but fungicides can be irritating to the skin and eyes. Inhalation of spray mist or dust from these pesticides may cause throat irritation, sneezing, and coughing. Chronic exposures to lower concentrations of fungicides can cause adverse health effects. Most cases of human fungicide poisonings have been from consumption of seed grain. To prevent these types of poisonings, fungicide treatment now includes a brightly colored dye to clearly indicate that the seed has been treated. Table summarizes the signs and symptoms of acute exposures to commonly used fungicides.

Table: Signs and symptoms of acute exposure for several fungicide active ingredients.

Active Ingredient	Brand Name	Signs and Symptoms.
Azoxystrobin	Abound, Quadris	Irritating to skin, eyes, respiratory tract.
Captan	Captol, Orthocide	Irritating to skin, eyes, respiratory tract.
Chlorothalonil	Bravo, Daconil	Irritation to skin, mucous membranes of the eye, respiratory tract. Allergic contact dermatitis.
Copper Compounds	Bordeaux mixture, Copper sulfate	Irritating to skin, eyes, respiratory tract. Salts are corrosive to mucous membranes and cornea Metallic taste, nausea, vomiting, intestinal pain.
Mancozeb	Dithane M-45, Manzate 200	Irritating to skin, eyes, respiratory tract.
Maneb	Dithane M-22, Manzate	Irritating to skin, eyes, respiratory tract. Skin disease in occupationally exposed individuals.
Pentachloronitrobenzene	PCNB, Terraclor	Allergic reactions.
Sulfur	Cosan, Thiolux	Irritating to skin, eyes, respiratory tract. Breath odor of rotten eggs. Diarrhea.. Irritant dermatitis in occupationally exposed individuals.
Thiram	Polyram-Ultra, Spotrete-F	Irritating to skin, eyes, respiratory mucous membranes.
Ziram	Cuman, Vancide	Irritating to skin, eyes, respiratory tract Prolonged inhalation causes neural and visual disturbances.

Herbicides

In general, herbicides have a low acute toxicity to humans because the physiology of plants is so different than that of humans. However, there are exceptions; many can be dermal irritants since they are often strong acids, amines, esters, and phenols. Inhalation of spray mist may cause coughing and a burning sensation in the nasal passages and chest. Prolonged inhalation sometimes causes dizziness. Ingestion will usually cause vomiting, a burning sensation in the stomach, diarrhea, and muscle twitching. Table summarizes the signs and symptoms of acute exposures to commonly used herbicides.

Table: Signs and symptoms of acute exposure for several herbicide active ingredients.

Active Ingredient	Brand Name	Signs and Symptoms
2,4-dichlorophenoxyacetic acid	2,4-D, Barrage	Irritating to skin, mucous membranes. Vomiting, headache, diarrhea, confusion. Bizarre or aggressive behavior. Muscle weakness in occupationally exposed individuals.
Acetochlor	Harness, Surpass	Irritating to skin, eyes, respiratory tract.
Atrazine	Aatrex, Atranex	Irritating to skin, eyes, respiratory tract. Abdominal pain, diarrhea, vomiting. Eye irritation, irritation of mucous membranes, skin reactions.
Dicamba	Banvel, Metambane	Irritating to skin, respiratory tract. Loss of appetite (anorexia), vomiting, muscle weakness, slowed heart rate, shortness of breath. Central nervous system effects.
Glyphosate	Rodeo, Roundup	Irritating to skin, eyes, respiratory tract.

Mecoprop	Kilporp, MCPP	Irritating to skin, mucous membranes. Vomiting, headache, diarrhea, confusion. Bizarre or aggressive behavior. Muscle weakness in occupationally exposed individuals.
Metolachlor	Bicep, Dual	Irritating to skin, eyes.
Paraquat	Gramoxone	Burning in mouth, throat, chest, upper abdomen. Diarrhea. Giddiness, headache, fever, lethargy. Dry, cracked hands, ulceration of skin.
Pendimethalin	Prowl, Stomp	Irritating to skin, eyes, respiratory tract.
Propanil	Propanex, Stampede	Irritating to skin, eyes, respiratory tract.

Insecticides

Insecticides cause the greatest number of pesticide poisonings in the United States. The most serious pesticide poisonings usually result from acute exposure to organophosphate and carbamate insecticides. Organophosphate insecticides include chlorpyrifos, diazinon, dimethoate, disulfoton, malathion, methyl parathion, and ethyl parathion. The carbamate compounds include carbaryl, carbofuran, methomyl, and oxamyl. Organophosphates and carbamates inhibit the enzyme cholinesterase, causing a disruption of the nervous system. All life forms with cholinesterase in their nervous system, such as insects, fish, birds, humans, and other mammals, can be poisoned by these chemicals.

Table: The signs and symptoms from acute exposures to commonly used insecticides.

Active Ingredient	Brand Name	Signs and Symptoms
Acephate (organophosphate)	Orthene	Headache, excessive salivation and tearing, muscle twitching, nausea, diarrhea. Respiratory depression, seizures, loss of consciousness. Pinpoint pupils.
Aldicarb (N-methyl carbamate)	Temik	Malaise, muscle weakness, dizziness, sweating. Headache, salivation, nausea, vomiting, abdominal pain, diarrhea. Nervous system depression, pulmonary edema in serious cases.
Carbaryl (N-methyl carbamate)	Sevin	Malaise, muscle weakness, dizziness, sweating. Headache, salivation, nausea, vomiting, abdominal pain, diarrhea. Nervous system depression, pulmonary edema in serious cases.
Chlorpyrifos (organophosphate)	Dursban	Headache, excessive salivation and tearing, muscle twitching, nausea, diarrhea. Respiratory depression, seizures, loss of consciousness. Pinpoint pupils.
Endosulfan (organochlorine)	Thiodan	Itching, burning, tingling of skin. Headache, dizziness, nausea, vomiting, lack of coordination, tremor, mental confusion. Seizures, respiratory depression, coma.
Malathion (organophosphate)	Cythion	Headache, excessive salivation and tearing, muscle twitching, nausea, diarrhea. Respiratory depression, seizures, loss of consciousness. Pinpoint pupils.
Methyl Parathion (organophosphate)	Penncap-M	Headache, excessive salivation and tearing, muscle twitching, nausea, diarrhea. Respiratory depression, seizures, loss of consciousness. Pinpoint pupils.

Phosmet (organophosphate)	Imidan	Headache, excessive salivation and tearing, muscle twitching, nausea, diarrhea. Respiratory depression, seizures, loss of consciousness. Pinpoint pupils.
Pyrethrins (natural origin)		Irritating to skin and upper respiratory tract. Contact dermatitis and allergic reactions - asthma.
Pyrethroids (synthetic pyrethrin)	Cypermethrin, permethrin	Abnormal facial sensation, dizziness, salivation, headache, fatigue, vomiting, diarrhea. Irritability to sounds or touch. Seizures, numbness.

To understand how the organophosphate and carbamate insecticides affect the nervous system, one needs to understand how the nervous system actually works. The nervous system, which includes the brain, is the most complex system in the body consist- ing of millions of cells that make up a communications system within the organism. Messages or electrical impulses (stimuli) travel along this complex network of cells. Nerve cells or neurons do not physically touch each other; rather there is a gap or synapse between cells. The impulses must cross or "bridge" the synapse between nerve cells in order to keep the message moving along the entire network.

When an impulse reaches the synapse, the chemical acetylcholine is released to carry the message on to the next cell. Acetylcholine is the primary chemical responsible for the transmission of nerve impulses across the synapse of two neurons. After the impulse is transmitted across the synapse, the acetylcholine is broken down by the enzyme cholinesterase. Once this occurs, the synapse is "cleared" and ready to receive a new transmission.

Organophosphate and carbamate insecticides inhibit the activity of cholinesterase, resulting in a buildup of acetylcholine in the body. An increase in acetylcholine results in the uncontrolled flow of nerve transmissions between nerve cells. The nervous system becomes "poisoned"; the accumulation of acetylcholine causes the continual transmission of impulses across the synapses.

The effects of organophosphate or carbamate poisoning can result in both systemic and topical symptoms. Direct exposure of the eye, for example, can cause topical symptoms such as constriction of the pupils, blurry vision, an eyebrow headache, and severe irritation and reddening of the eyes. Symptoms and signs of systemic poisonings are almost entirely due to the accumulation of acetylcholine at the nerve endings.

The onset of symptoms depends on the route of entry and the severity of the poisoning. Gastric symptoms such as stomach cramps, nausea, vomiting, and diarrhea appear early if the material has been ingested. Similarly, salivation, headache, dizziness, and excessive secretions that cause breathing difficulties are initial symptoms if the material has been inhaled. Involvement of the respiratory muscles can result in respiratory failure. Stomach, intestinal, and respiratory symptoms usually appear at the same time if the pesticide is absorbed through the skin. In children, the first symptom of poisoning may be a convulsion.

In advanced poisonings, the victim is pale, sweating, and frothing at the mouth. The pupils are constricted and unresponsive to light. Other symptoms include changes in heart rate, muscle weakness, mental confusion, convulsions, and coma. The victim may die if not treated.

Cholinesterase Testing

Those who regularly work with organophosphates and carbamates should consider having periodic cholinesterase tests. The blood cholinesterase test measures the effect of exposure to organophosphate and carbamate insecticides. Since cholinesterase levels can vary considerably among individuals, a "baseline" must be established for each person. In fact, a small percentage of the population has a genetically determined low level of cholinesterase. Even minimal exposure to cholinesterase inhibitors can present a substantial risk to these people. Baseline testing should always be done during the time of year when pesticides are not being used, or at least 30 days from the most recent exposure. Establishing a baseline value often requires two tests performed at least 72 hours apart but within 14 days of each other. If the test results differ by as much as 20 percent, a third test is often recommended.

Cholinesterase tests can be repeated during times when organophosphate and carbamate insecticides are being used and then compared with the baseline level. The purpose of routine cholinesterase monitoring is to enable a physician to recognize the occurrence of excessive exposure to organophosphates and carbamates. If a laboratory test shows a cholinesterase drop of 30 percent below the established baseline, the worker should be retested immediately. If a second test confirms the drop in cholinesterase, the pesticide handler or agricultural worker should be removed from further contact with organophosphate and carbamate insecticides until cholinesterase levels return to the pre-exposure baseline range. Your primary care physician can help to establish the frequency of this testing program.

Exposure and Preventative Measures

The hazard or risk involved with using a pesticide depends on both the toxicity of the product and the amount of exposure to the product (Hazard = Toxicity × Exposure). Ideally, use a low-toxicity product when possible, but even they can be harmful if your exposure level is high. However, regardless of the product's toxicity, if the exposure level is low, then the hazard will also be low. To reduce the possibility of exposure and to protect your health, always wear the personal protective equipment (PPE) as indicated on the product label. The following are general PPE guidelines to protect against the four routes of entry.

Dermal

More than 95 percent of all exposures are dermal. Dermal absorption may occur as the result of a splash, spill, or drift or when cleaning or repairing equipment. Wear unlined, chemical-resistant gloves to eliminate most dermal exposures. Minimum dermal protection for most pesticides consists of a long-sleeved shirt, long trousers, gloves, and proper footwear. For extra precaution, consider wearing coveralls, a waterproof hat, and unlined rubber boots. Additionally, wearing a liquid-proof apron or rain suit is recommended when mixing and pouring concentrates or when using highly toxic products.

Inhalation

For many toxic chemicals, the respiratory (breathing) system is the quickest and most direct route of entry into the circulatory system. Respiratory protection is especially important when pesticide powders, dusts, gases, vapors, or small spray droplets can be inhaled.

Use the respirator as designed for its intended use, and always follow the manufacturer's instructions. Select only equipment approved by the National Institute of Occupational Safety and Health (NIOSH) and the Mine Safety and Health Administration (MSHA).

Oral

Accidental oral exposure most frequently occurs when pesticides have been taken from the original container and put into an unlabeled bottle or food container. Unfortunately, children are the most common victims in these situations. Store pesticides only in their original containers, and keep the original label attached to the container. Store pesticides only in their original containers and keep the original label attached to the container. Store in a locked cabinet and on a high shelf to keep out of the reach of children. Never use your mouth to clear a spray line or to siphon a pesticide from a tank or container. After handling or working with pesticides, wash your hands and face thoroughly with soap and water before eating, drinking, or smoking.

Eyes

Eyes are very sensitive to many pesticides and, considering their size, are able to absorb large amounts of chemical. Serious eye exposure can result from a splash, spill, or drift or by rubbing the eyes with contaminated hands or clothing. Tight-fitting chemical splash goggles or a full-face shield should be worn if there is any chance of getting pesticides in the eyes, especially when pouring or mixing concentrates and handling dusts. When pouring from a container, keep the container below eye level to avoid splashing or spilling chemicals on your face or protective clothing.

EFFECTS OF PESTICIDES IN FOOD

Pesticides, which are any substance intended to prevent or destroy pests, are used to protect food from bacteria, weeds, mold, insects and rodents. According to the Environmental Protection Agency, pesticides can be harmful to people, animals or the environment because they are designed to kill or harm living organisms. Because of this, pesticide residue on the foods you eat can have an effect on your health. Though the government regulates pesticide use, residues are still found in our food supply.

Pesticides are intended to protect food from insects and rodents.

Children who are frequently exposed to a small amount of organophosphates, a pesticide found on commercially grown fruit and vegetables, are more likely to have attention deficit hyperactivity disorder than children who have been exposed less often. Exposure to these pesticides has also been linked to behavior and learning problems in children.

Nervous System

Organophsphates, frequently found on peaches, also affect the nervous system.

Organophosphates have also been shown to affect the nervous system. Signs of an affected nervous system include excess salivation, stomach pain, vomiting, constipation and diarrhea. The most pesticides are found on celery, peaches, berries, apples, peppers, greens, grapes and potatoes. Eating these commercially grown foods more frequently might increase exposure.

Breast Cancer

Pesticides can be linked to some cases of breast cancer.

According to Environmental Health Perspectives, a journal published by the National Institute of Environmental Health Sciences, growing evidence shows that pesticides found in commercially grown food can be linked to some cases of breast cancer. The risk increases with combined with other factors, including lifestyle, diet and genes.

Weakened Immune System

Pesticides may alter the immune system.

According to the Global Healing Center, several studies have shown that pesticides alter the immune system in animals and make them more susceptible to disease. Pesticides have been found to reduce the numbers of white blood cells and disease-fighting lymphocytes, making their bodies unable to kill bacteria and viruses. According to the GHC, they also affect the development of the spleen and thymus and spleen, two immune organs. However, studies on humans have been inconclusive.

PESTICIDE DRIFT

Pesticide drift is the movement of a spray solution from the intended target to a place where it is not wanted, or the movement of spray droplets or pesticide vapors out of the sprayed area. In particular, herbicide spray drift can damage shelterbelts, garden and ornamental plants, cause water pollution, and damage non-susceptible crops in a vulnerable growth stage (2,4-D drift on wheat in the flowering or seedling stage, for example). Herbicide spray drift can cause non-uniform application in a field, with possible crop damage and poor weed control. In addition, insecticide spray drift can damage beneficial insect populations especially bees and natural predators of Montana agricultural pests. Drift is also costly from a financial standpoint. If only 50 percent of an applied solution makes it to the target, then you have wasted 50 percent of what you have applied. In all the above cases, the pesticide becomes an environmental pollutant, injuring susceptible plants, contaminating water, wildlife and even humans. Sadly, the majority of pesticide spray drift problems involve mistakes that could have been avoided by the applicator.

The effects of a spray solution once it leaves the nozzle are the responsibility of the applicator.

Given the threat of lawsuits, fines and human health issues, it is inconceivable that pesticide drift continues to be a problem. Yet every year there are reports of crop damage, damage to landscape plants, livestock contamination and human health problems. All contributable to drift, usually as a result of applicator error.

Remember when you spray a site you represent all pesticide applicators and a good portion of Montana agriculture. Irresponsibility in the field will only anger those whose only opinion of agriculture is the one you have given them.

If drift problems become too frequent or too serious, some of our most useful pesticides could be withdrawn or their use severely restricted.

In particular, some herbicides are much more of a problem than others. Some herbicides contain the active ingredient (a.i.) glyphosate (Roundup) which is non-selective - it kills any plant it touches. On the other hand, the herbicide 2,4-D is selective and kills or damages only broadleaf plants. Because ester formulations of 2,4-D and occasionally Banvel can vaporize under high temperature and move to off target plant species, it is best to minimize applications during windy days or extremely hot weather to avoid problems.

Types of Drift

The best way to reduce drift is to understand the factors which cause it, most of which can be controlled by the applicator. It begins with knowing what drift is and how drift is best managed. There are two kinds of drift:

- Particle drift is off-target movement of the spray particles.

- Vapor drift is the volatilization of the pesticide molecules and their movement off target.

Factors of Drift

- Wind Speed: When the wind speed was doubled, there was almost a 70% increase in drift when the readings were taken 90 feet downwind from the sprayer. Spray when the wind speed is 10 mph wind or less.

- Boom height: When the boom height was increased from 18 to 36 inches the amount of drift increased 350% at 90 feet downwind.

- Distance downwind: If the distance downwind is doubled, the amount of drift decreases five-fold. If the distance downwind goes from 100 to 200 feet, you have only 20% as much drift at 200 feet as at 100 feet and if the distance goes to 400 feet, you only have 4% of the drift you had at 100 feet. Check wind direction and speed when starting to spray a field. You may want to start spraying one side of the field when the wind is lower. Also it may be necessary to only spray part of a field because of wind speed, wind direction and distance to susceptible vegetation. The rest of the field can be sprayed when conditions change.

Other important factors that must be considered in drift management are spray pressure, nozzle size, nozzle orientation, operating speed, air temperature, relative humidity, shields on sprayers and nozzles, application rate and instructions from the manufacturer of the spray product.

Drift Reduction Nozzles

Many new nozzles for reducing drift are now available. Many of these use a pre-orifice which controls the flow rate. The exit orifice controls the pattern formation. The result is larger spray droplets which are less susceptible to drift.

Control of volunteer wheat and triazine-resistant kochia nine days after treatment with paraquat plus atrazine at 0.31 plus 0.5 lb/A with a nonionic surfactant 0.25% v/v.				
Trt.	Nozzle[a]	Droplet Size[b]	Volunteer Wheat %	TR-Kochia %
1	XR11005	Medium	97 a	97 a
2	DG11005	Coarse	96 a	95 a
3	TF-VS2.5	Extremely Coarse	79 c	80 c

[a]All nozzles on 30-inch spacing operated at 30 psi, 8.6 mph, delivering 10 gpa.

[b]Very fine <153, fine 154-241, medium 242-358, coarse 359-451, very coarse 452-740, and extremely coarse >741.

Also, some of these nozzles can be used over a wider pressure range, which produces large droplets at low pressure and small droplets at high pressures. The ability of these nozzles to produce good spray patterns over a wide pressure range makes them an excellent choice to use with rate controllers which control the application rate by pressure changes. These drift reduction nozzles can still have a problem with drift especially when the sprayer speed is increased and pressure therefore is increased, resulting in smaller spray droplets. At slow speeds the spray droplets may be too large for good coverage. There are advantages and disadvantages with nozzles which produce large spray droplets. For most postemergence contact, herbicide coverage is important. Nozzles which induce air are also available which use a pre-orifice. Early research work with these nozzles on drift reduction show most of the benefit as to particle size is from the pre-orifice.

Remember, the environment and human safety are the top priority of any activity. There are no excuses for mishandled herbicides when human safety is the issue. With the larger number of people coming into contact with agriculture, we need to be sensitive to their lack of knowledge of agricultural issues. Understanding drift and knowing how others have learned to manage it will help most producers avoid problems. Bottom line, we are responsible for the injury we cause and are accountable for it. Keeping pesticides confined to the target area is an on-going problem. We can't blame our neighbors if they get upset because our pesticides drift onto their property.

Low-volatile (LV) esters are not really low volatile. Indeed, they are less volatile than the old high-volatile ones (butyl esters), but the LV esters are still considerably more volatile than amines. LV esters are more susceptible to movement because gases can move farther than spray droplets, and can come off of previously sprayed plants or soil. Choose the amine form if there are susceptible plants in the area.

Even nonvolatile chemicals can drift. Small spray droplets can move considerable distances in some weather conditions.

Keep spray droplets as large as practical. For most pesticide usage, especially with 2,4-D type herbicides, a minimum size of 0.2 gal/min (for example, Spraying Systems 8002) flat fan nozzle tips and a maximum of 30 psi pressure are sufficient for good coverage. Smaller nozzle tips or higher pressure can produce too many "fines," or small droplets, which can easily move laterally to non-target areas. Some herbicide labels call for application at higher pressure. Apply these products with extra caution. Flood-type nozzles can reduce spray drift by producing larger droplets at low pressure. They produce a less precise pattern than flat fan nozzles, but in many situations they are satisfactory. Consider using a new generation of flat fan nozzles designed for lower pressures when the precision of the flat fan is required.

A windscreen may reduce drift. A windscreen around the boom and reaching near the sprayed surface may reduce drift. To avoid a chimney effect, place windscreens above the boom. Because the spray pattern cannot be seen by the operator, sprayers can be equipped with tip monitors to detect plugged nozzle tips.

A drift-control adjuvant, such as Nalcotrol, may help reduce the production of small droplets, thereby reducing drift.

Proper timing of herbicide application can help avoid damage to nearby plants. For example, grapes are readily injured by 2,4-D-type herbicides (such as Crossbow). The greatest damage to fruit production seems to be when drift occurs after the fruiting cluster has emerged but before bloom, generally mid- to late-May. Always avoid drift, but in areas where grapes are grown, not spraying during sensitive stages may be the safest approach. Observe the same principle with other sensitive plants in your area.

Use wide-angle nozzle tips to keep the boom low. Research indicates doubling the boom height quadruples drift. Of course, the drift potential from aerial application is considerably higher than from ground application.

The biggest single weather factor involved in drift is wind. Even relatively light breezes can carry small droplets a long distance. Generally, spraying early in the morning is preferable to afternoon or evening. If you are spraying near sensitive crops, limit your applications to times when winds do not exceed 5 mph. Spraying when slight winds are away from sensitive crops may be safer than spraying when the air is calm.

Consider not spraying areas nearest to sensitive crops. Leave a buffer zone. Do not apply pesticide to dusty soil which might later be carried on winds to sensitive crops or aquatic areas. Do not apply pesticides to areas where treated soil is likely to be carried by water to where sensitive crops are grown.

Avoiding chemical trespass is the responsibility of each pesticide user. This requires intelligent management and great care. Pesticide labels include useful information about any special characteristics of the product related to off-target movement.

Use pesticides safely:

- Wear protective clothing and safety devices as recommended on the label.

- Bathe or shower after each use.

- Read the pesticide label - even if you've used the pesticide before.

- Follow closely the instructions on the label (and any other directions you have).

- Be cautious when you apply pesticides.

- Know your legal responsibility as a pesticide applicator. You may be liable for injury or damage resulting from pesticide use.

Because all nozzles produce a range of droplet sizes, the small, drift-prone particles cannot be completely eliminated, but drift can be reduced and kept within reasonable limits:

- Use adequate amounts of carrier. This means larger nozzles, which in turn usually produces larger droplets. Although this will increase the number of refills, the added carrier improves coverage and usually increases the effectiveness of the chemicals. Smaller droplets will be produced with lower spray volumes, resulting in a greater drift hazard.

- Avoid using high pressure. Higher pressures create fine droplets; 40 PSI should be considered the maximum for conventional broadcast spraying.

- Use a drift-reducing nozzle where practical. They produce larger droplets and operate at lower pressure than the equivalent flat-fan nozzle.

- Many drift-reducing spray additives which can be used with regular spray equipment are available today.

- Use wide angle nozzles and keep the boom stable and as close to the crop as possible.

- Spray when wind speeds are less than 10 mph and when wind direction is away from sensitive crops.

- Do not spray when the air is completely calm or when an inversion exists.

- Use a shielded spray boom when wind conditions exceed prime pesticide application conditions.

Drift Prevention and Sprayer Calibration

The amount of chemical solution applied per acre depends on forward speed, system pressure, size of nozzle, and nozzle spacing on the boom. A change in any one of these will change the application rate. The application rate is usually expressed as Gallons Per Acre or GPA. The forward speed and pressure must be adjusted correctly to set the sprayer for any given rate per acre. The nozzle size should be changed to make a large change in application rates, and all nozzles should discharge an equal amount of spray. If any of these adjustments are incorrect, poor results will be obtained.The first thing to do with sprayer calibration is select the type and size nozzle for your spraying job. You can base the nozzle type decision on spraying conditions and guidelines as recommended in the following tables. Once you've selected the type of nozzle, the next step is to calculate the nozzle size. Nozzle selection should not be based on "Gallons Per Acre" or GPA as advertised by some manufacturers. Certain GPA standards are usually required by some pesticide labels. A nozzle that is identified as a 10-gallon nozzle will deliver this amount per acre only for one condition, such as when the nozzle spacing is 20 inches on the boom, the sprayer is traveling at 4 mph and the boom pressure is 30 psi. If the spacing, speed or pressure varies from these set values, the nozzle will not deliver the specified gallons per acre. If the pressure is too high, then small droplets will be produced resulting in drift. Choice of nozzle size should be based on a gallons-per-minute calculation rather than a gallons-per-acre calculation. Basing calculations on gpm allows the operator to make the spraying decisions based on the crop and field conditions.

	Extended Range Flat Fan	Standard Flat Fan	Drift Guard Flat Fan	Twin Flat Fan	Turbo Flood Wide Angle	Wide Angle Full Core	Flood Nozzle Wide Angle	Raindrop Hollow Cone
Herbicides								
Soil-incorporated	Good		Very Good		Very Good	Very Good	Good	Good
Pre-emerge	Very Good (at low pressure)	Good	Very Good		Very Good	Very Good		Good
Post-emerge Contact	Good	Good		Very Good				
Post-emerge Systemic	Very Good (at low pressure)	Good	Very Good		Very Good			Good
Fungicides								
Contact	Very Good	Good						
Systemic	Very Good (at low pressure)		Very Good		Very Good			
Insecticides								
Contact	Good	Good		Very Good				
Systemic	Very Good (at low pressure)		Very Good		Very Good			

	Even Flat Fan	Twin Even Flat Fan	Hollow Cone	Full Cone	Disc and Core Cone
Herbicides					
Pre-emerge	Very Good	Good		Good	
Post-emerge Contact	Good	Very Good	Very Good		
Post-emerge Systemic	Very Good	Good			
Fungicides					
Contact	Good		Good		Very Good
Systemic	Very Good				Good
Insecticides					
Contact		Very Good	Very Good		Very Good
Systemic	Very Good				Good
Growth Regulations	Good			Very Good	

PESTICIDE POISONING

A pesticide poisoning occurs when chemicals intended to control a pest affect non-target organisms such as humans, wildlife, or bees. There are three types of pesticide poisoning. The first of the three is a single and short-term very high level of exposure which can be experienced by individuals who commit suicide, as well as pesticide formulators. The second type of poisoning is long-term high-level exposure, which can occur in pesticide formulators and manufacturers. The third type of poisoning is a long-term low-level exposure, which individuals are exposed to from sources such as pesticide residues in food as well as contact with pesticide residues in the air, water, soil, sediment, food materials, plants and animals.

In developing countries, such as Sri Lanka, pesticide poisonings from short-term very high level of exposure (acute poisoning) is the most worrisome type of poisoning. However, in developed countries, such as Canada, it is the complete opposite: acute pesticide poisoning is controlled, thus making the main issue long-term low-level exposure of pesticides.

Cause

The most common exposure scenarios for pesticide-poisoning cases are accidental or suicidal poisonings, occupational exposure, by-stander exposure to off-target drift, and the general public who are exposed through environmental contamination.

Accidental or Suicidal

Self-poisoning with agricultural pesticides represents a major hidden public health problem accounting for approximately one-third of all suicides worldwide. It is one of the most common forms of self-injury in the Global South. The World Health Organization estimates that 300,000 people die from self-harm each year in the Asia-Pacific region alone. Most cases of intentional pesticide poisoning appear to be impulsive acts undertaken during stressful events, and the availability of pesticides strongly influences the incidence of self poisoning. Pesticides are the agents most frequently used by farmers and students in India to commit suicide.

Occupational

Pesticide poisoning is an important occupational health issue because pesticides are used in a large number of industries, which puts many different categories of workers at risk. Extensive use puts agricultural workers in particular at increased risk for pesticide illnesses. Exposure can occur through inhalation of pesticide fumes, and often occurs in settings including greenhouse spraying operations and other closed environments like tractor cabs or while operating rotary fan mist sprayers in facilities or locations with poor ventilation systems. Workers in other industries are at risk for exposure as well. For example, commercial availability of pesticides in stores puts retail workers at risk for exposure and illness when they handle pesticide products. The ubiquity of pesticides puts emergency responders such as fire-fighters and police officers at risk, because they are often the first responders to emergency events and may be unaware of the presence of a poisoning hazard. The process of aircraft disinsection, in which pesticides are used on inbound international flights for insect and disease control, can also make flight attendants sick.

Different job functions can lead to different levels of exposure. Most occupational exposures are caused by absorption through exposed skin such as the face, hands, forearms, neck, and chest. This exposure is sometimes enhanced by inhalation in settings including spraying operations in greenhouses and other closed environments, tractor cabs, and the operation of rotary fan mist sprayers.

Residential

When thinking of pesticide poisoning, one does not take into consideration the contribution that is made of their own household. The majority of households in Canada use pesticides while taking part in activities such as gardening. In Canada 96 percent of households report having a lawn or a garden. 56 percent of the households who have a lawn or a garden utilize fertilizer or pesticide. This form of pesticide use may contribute to the third type of poisoning, which is caused by long-term low-level exposure. As mentioned before, long-term low-level exposure affects individuals from sources such as pesticide residues in food as well as contact with pesticide residues in the air, water, soil, sediment, food materials, plants and animals.

Pathophysiology

Organochlorines

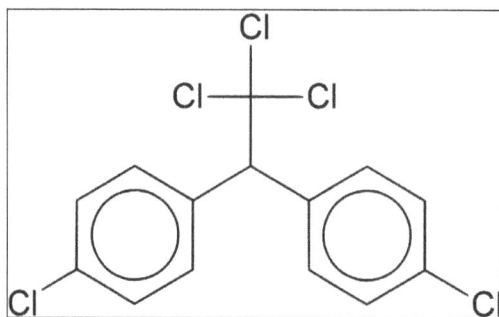

DDT, an organochlorine.

The organochlorine pesticides, like DDT, aldrin, and dieldrin, are extremely persistent and accumulate in fatty tissue. Through the process of bioaccumulation (lower amounts in the environment get magnified sequentially up the food chain), large amounts of organochlorines can accumulate in top species like humans. There is substantial evidence to suggest that DDT, and its metabolite DDE, act as endocrine disruptors, interfering with hormonal function of estrogen, testosterone, and other steroid hormones.

Anticholinesterase Compounds

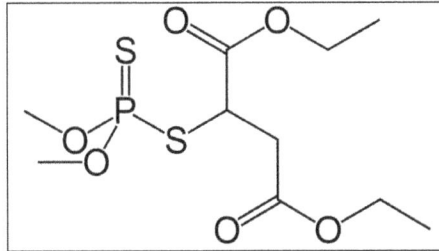

Malathion, an organophosphate anticholinesterase.

Cholinesterase-inhibiting pesticides, also known as organophosphates, carbamates, and anticholinesterases, are most commonly reported in occupationally related pesticide poisonings globally. Besides acute symptoms including cholinergic crisis, certain organophosphates have long been known to cause a delayed-onset toxicity to nerve cells, which is often irreversible. Several studies have shown persistent deficits in cognitive function in workers chronically exposed to pesticides.

Diagnosis

Most pesticide-related illnesses have signs and symptoms that are similar to common medical conditions, so a complete and detailed environmental and occupational history is essential for correctly diagnosing a pesticide poisoning. A few additional screening questions about the patient's work and home environment, in addition to a typical health questionnaire, can indicate whether there was a potential pesticide poisoning.

If one is regularly using carbamate and organophosphate pesticides, it is important to obtain a baseline cholinesterase test. Cholinesterase is an important enzyme of the nervous system, and these chemical groups kill pests and potentially injure or kill humans by inhibiting cholinesterase. If one has had a baseline test and later suspects a poisoning, one can identify the extent of the problem by comparison of the current cholinesterase level with the baseline level.

Prevention

Accidental poisonings can be avoided by proper labeling and storage of containers. When handling or applying pesticides, exposure can be significantly reduced by protecting certain parts of the body where the skin shows increased absorption, such as the scrotal region, underarms, face, scalp, and hands. Safety protocols to reduce exposure include the use of personal protective equipment, washing hands and exposed skin during as well as after work, changing clothes between work shifts, and having first aid trainings and protocols in place for workers.

Personal protective equipment for preventing pesticide exposure includes the use of a respirator, goggles, and protective clothing, which have all have been shown to reduce risk of developing pesticide-induced diseases when handling pesticides. A study found the risk of acute pesticide poisoning was reduced by 55% in farmers who adopted extra personal protective measures and were educated about both protective equipment and pesticide exposure risk. Exposure can be significantly reduced when handling or applying pesticides by protecting certain parts of the body where the skin shows increased absorption, such as the scrotal region, underarms, face, scalp, and hands. Using chemical-resistant gloves has been shown to reduce contamination by 33–86%.

Treatment

Specific treatments for acute pesticide poisoning are often dependent on the pesticide or class of pesticide responsible for the poisoning. However, there are basic management techniques that are applicable to most acute poisonings, including skin decontamination, airway protection, gastrointestinal decontamination, and seizure treatment.

Decontamination of the skin is performed while other life-saving measures are taking place. Clothing is removed, the patient is showered with soap and water, and the hair is shampooed to remove chemicals from the skin and hair. The eyes are flushed with water for 10–15 minutes. The patient is intubated and oxygen administered, if necessary. In more severe cases, pulmonary ventilation must sometimes be supported mechanically. Seizures are typically managed with lorazepam, phenytoin and phenobarbitol, or diazepam (particularly for organochlorine poisonings).

Gastric lavage is not recommended to be used routinely in pesticide poisoning management, as clinical benefit has not been confirmed in controlled studies; it is indicated only when the patient has ingested a potentially life-threatening amount of poison and presents within 60 minutes of ingestion. An orogastric tube is inserted and the stomach is flushed with saline to try to remove the poison. If the patient is neurologically impaired, a cuffed endotracheal tube inserted beforehand for airway protection. Studies of poison recovery at 60 minutes have shown recovery of 8%–32%. However, there is also evidence that lavage may flush the material into the small intestine, increasing absorption. Lavage is contra-indicated in cases of hydrocarbon ingestion.

Activated charcoal is sometimes administered as it has been shown to be successful with some pesticides. Studies have shown that it can reduce the amount absorbed if given within 60 minutes, though there is not enough data to determine if it is effective if time from ingestion is prolonged. Syrup of ipecac is not recommended for most pesticide poisonings because of potential interference with other antidotes and regurgitation increasing exposure of the esophagus and oral area to the pesticide.

Urinary alkalinisation has been used in acute poisonings from chlorophenoxy herbicides (such as 2,4-D, MCPA, 2,4,5-T and mecoprop); however, evidence to support its use is poor.

Epidemiology

Acute pesticide poisoning is a large-scale problem, especially in developing countries.

"Most estimates concerning the extent of acute pesticide poisoning have been based on data from hospital admissions which would include only the more serious cases. The latest estimate by a WHO task group indicates that there may be 1 million serious unintentional poisonings each year and in addition 2 million people hospitalized for suicide attempts with pesticides. This necessarily reflects only a fraction of the real problem. On the basis of a survey of self-reported minor poisoning carried out in the Asian region, it is estimated that there could be as many as 25 million agricultural workers in the developing world suffering an episode of poisoning each year." In Canada in 2007 more than 6000 cases of acute pesticide poisoning occurred.

Estimating the numbers of chronic poisonings worldwide is more difficult.

Society and Culture

Rachel Carson's 1962 environmental science book Silent Spring brought about the first major wave of public concern over the chronic effects of pesticides.

In other Animals

An obvious side effect of using a chemical meant to kill is that one is likely to kill more than just the desired organism. Contact with a sprayed plant or "weed" can have an effect upon local wildlife, most notably insects. A cause for concern is how pests, the reason for pesticide use, are building up a resistance. Phytophagous insects are able to build up this resistance because they are easily capable of evolutionary diversification and adaptation. The problem this presents is that in order to obtain the same desired effect of the pesticides they have to be made increasingly stronger as time goes on. Repercussions of the use of stronger pesticides on vegetation has a negative result on the surrounding environment, but also would contribute to consumers' long-term low-level exposure.

PESTICIDE TOXICITY TO BEES

Pesticides vary in their effects on bees. Contact pesticides are usually sprayed on plants and can kill bees when they crawl over sprayed surfaces of plants or other areas around it. Systemic pesticides, on the other hand, are usually incorporated into the soil or onto seeds and move up into the stem, leaves, nectar, and pollen of plants.

Of contact pesticides, dust and wettable powder pesticides tend to be more hazardous to bees than solutions or emulsifiable concentrates. When a bee comes in contact with pesticides while foraging, the bee may die immediately without returning to the hive. In this case, the queen bee, brood, and nurse bees are not contaminated and the colony survives. Alternatively, the bee may come into contact with an insecticide and transport it back to the colony in contaminated pollen or nectar or on its body, potentially causing widespread colony death.

Actual damage to bee populations is a function of toxicity and exposure of the compound, in combination with the mode of application. A systemic pesticide, which is incorporated into the soil or coated on seeds, may kill soil-dwelling insects, such as grubs or mole crickets as well as other insects, including bees, that are exposed to the leaves, fruits, pollen, and nectar of the treated plants.

Pesticides are linked to Colony Collapse Disorder and are now considered a main cause, and the toxic effects of Neonicotinoids on bees are confirmed. Currently, many studies are being conducted to further understand the toxic effects of pesticides on bees. Agencies such as the EPA and EFSA are making action plans to protect bee health in response to calls from scientists and the public to ban or limit the use of the pesticides with confirmed toxicity.

Classification

Insecticide toxicity is generally measured using acute contact toxicity values LD_{50} – the exposure level that causes 50% of the population exposed to die. Toxicity thresholds are generally set at:

- Highly toxic (acute LD50 < 2µg/bee).

- Moderately toxic (acute LD50 2 - 10.99µg/bee).

- Slightly toxic (acute LD50 11 - 100µg/bee).

- Nontoxic (acute LD50 > 100µg/bee) to adult bees.

Pesticide Toxicity

Acute Toxicity

The acute toxicity of pesticides on bees, which could be by contact or ingestion, is usually quantified by LD. Acute toxicity of pesticides causes a range of effects on bees, which can include agitation, vomiting, wing paralysis, arching of the abdomen similar to sting reflex, and uncoordinated movement. Some pesticides, including Neonicotinoids, are more toxic to bees and cause acute symptoms with lower doses compared to older classes of insecticides. Acute toxicity may depend on the mode of exposure, for instance, many pesticides cause toxic effects by contact while Neonicotinoids are more toxic when consumed orally. The acute toxicity, although more lethal, is less common than sub-lethal toxicity or cumulative effects.

Sub-lethal and Chronic Effects

Field exposure of bees to pesticides, especially with relation to neonicotinoids, is most commonly sub-lethal. Sub-lethal effects to honey bees are of major concern and include behavioral disruptions such as disorientation, thermoregulation, reduced foraging, decreased flight and locomotion abilities, impaired memory and learning, phototaxis (response to light), and a shift in communication behaviors. Additional sub-lethal effects may include compromised immunity of bees and delayed development.

Neonicotinoids are especially likely to cause cumulative effects on bees due to their mechanism of function as this pesticide group works by binding to nicotinic acetylcholine receptors in the brains of the insects, and such receptors are particularly abundant in bees. Over-accumulation of acetylcholine results in paralysis and death.

Colony Collapse Disorder

Colony collapse disorder is a syndrome that is characterized by the sudden loss of adult bees from the hive. Many possible explanations for it have been proposed, but no one primary cause has been found. The US Department of Agriculture has indicated in a report to Congress that a combination of factors may be causing colony collapse disorder, including pesticides, pathogens, and parasites, all of which have been found at high levels in affected bee hives.

The development of a bee from egg to adult takes about three weeks. The queens daily laying rate will decline if contaminated materials are brought back to the hive such as pesticides 31.6% of exposed honey bees will fail to return to their colony every day while the rest will bring back contaminated pollen which in turn will not only affect the worker bees but also the queen. As a consequence there will be an upset in colony dynamics.

Colony Collapse Disorder has more implication than the extinction of one bee species; the disappearance of honeybees can cause catastrophic health and financial impacts. Honeybee pollination has an estimated value of more than $14 billion annually to the United States agriculture. Honeybees are required for pollinating many crops, which range from nuts to vegetables and fruits, that are necessary for human and animal diet.

The EPA updated their guidance for assessing pesticide risks to honeybees in 2014. For the EPA, when certain pesticide use patterns or triggers are met, current test requirements include the honey bee acute contact toxicity test, the honey bee toxicity of residues on foliage test, and field testing for pollinators. EPA guidelines have not been developed for chronic or acute oral toxicity to adult or larval honey bees. On the other hand, the PMRA (Pest Management Regulatory Agency) requires both acute oral and contact honey bee adult toxicity studies when there is potential for exposure for insect pollinators. Primary measurement endpoint derived from the acute oral and acute contact toxicity studies is the median lethal dose for 50% of the organisms tested (i.e., LD_{50}), and if any biological effects and abnormal responses appear, including sub-lethal effects, other than the mortality, it should be reported.

The EPA's testing requirements do not account for sub-lethal effects to bees or effects on brood or larvae. Their testing requirements are also not designed to determine effects in bees from exposure to systemic pesticides. With colony collapse disorder, whole hive tests in the field are needed in order to determine the effects of a pesticide on bee colonies. To date, there are very few scientifically valid whole hive studies that can be used to determine the effects of pesticides on bee colonies because the interpretation of such whole-colony effects studies is very complex and relies on comprehensive considerations of whether adverse effects are likely to occur at the colony level.

A March 2012 study conducted in Europe, in which minuscule electronic localization devices were fixed on bees, has shown that, even with very low levels of pesticide in the bee's diet, a high proportion of bees (more than one third) suffers from orientation disorder and is unable to come back to the hive. The pesticide concentration was order of magnitudes smaller than the lethal dose used in the pesticide's current use. The pesticide under study, brand-named "Cruiser" in Europe (thiamethoxam, a neonicotinoid insecticide), although allowed in France by annually renewed exceptional authorization, could be banned in the coming years by the European Commission.

Based on a risks to bee health as identified by EFSA, in April 2013 the EU decided to restrict thiamethoxam, clothianidin, and imidacloprid. The UK voted against the ban saying it would harm food production. Agrochemical companies Syngenta and Bayer CropScience both began legal proceedings to object to the ban. It is their position that there is no science that implicates their pesticide products.

Bee Kill Rate Per Hive

The kill rate of bees in a single bee hive can be classified as:

- < 100 bees per day - normal die off rate.

- 200-400 bees per day - low kill.

- 500-900 bees per day - moderate kill.

- > 1000 bees per day - High Kill.

Pesticides Formulations

Pesticides come in different formulations:

- Dusts (D).

- Wettable powders (WP).

- Soluble powders (SP).

- Emulsifiable concentrates (EC).

- Solutions (LS).

- Granulars (G).

EPA Proposal to Protect Bees from Acutely Toxic Pesticides

The EPA is proposing to prohibit the application of certain pesticides and herbicides known toxic to bees during pollination periods when crops are in bloom. Growers routinely contract with honeybee keepers to bring in bees to pollinate their crops that require insect pollination. Bees are typically present during the period the crops are in bloom. Application of pesticides during this period can significantly affect the health of bees. These restrictions are expected to reduce the likelihood of high levels of pesticide exposure and mortality for bees providing pollination services. Moreover, the EPA believes these additional measures to protect bees providing pollination services will protect other pollinators as well.

The proposed restrictions would apply to all products that have liquid or dust formulations as applied, foliar use (applying pesticides directly to crop leaves) directions for use on crops, and active ingredients that have been determined via testing to have high toxicity for bees (less than 11 micrograms per bee). These restrictions would not replace already existing more restrictive, chemical-specific, and bee-protective provisions. Additionally, the proposed label restrictions would not apply to applications made in support of a government-declared public health response, such as use for wide area mosquito control. There would be no other exceptions to these proposed restrictions.

General Measures to Prevent Pesticides Bee Kills

Application of Pesticides at Evening or Night

Avoiding pesticide application directly to blooming flowers as much as possible can help limit the exposure of honeybees to toxic materials as honeybees are attracted to all types of blooming flowers. If blooming flowers must be sprayed with pesticides for any reason, they should be sprayed in the evening or night hours as bees are not in the field at that time. Usual foraging hours of honeybees are when the temperature is above 55-60 °F during the daytime, and by the evening, the bees return to the hives.

References

- Bertolote, J. M.; Fleischmann, A.; Eddleston, M.; Gunnell, D. "Deaths from pesticide poisoning: a global response". The British Journal of Psychiatry. Archived from the original on 4 March 2015. Retrieved 26 April 2018

- What-is-the-environmental-impact-of-pesticides: worldatlas.com, Retrieved 7 July, 2019

- Reeves, K. S.; Schafer, K. S. (2003). "Greater risks, fewer rights: U.S. Farmworkers and pesticides". International Journal of Occupational and Environmental Health. 9 (1): 30–39. Doi:10.1179/107735203800328858. PMID 12749629

- Potential-health-effects-of-pesticides: psu.edu, Retrieved 8 June, 2019

- Rabesandratana, Tania. "Pesticidemakers Challenge E.U. Neonicotinoid Ban in Court". Science. Retrieved July 4, 2017

- The-effects-of-pesticides-in-food: ivestrong.com, Retrieved 9 August, 2019

- Tosi, Simone; Burgio, Giovanni; Nieh, James C (2017). "A common neonicotinoid pesticide, thiamethoxam, impairs honey bee flight ability". Scientific Reports. 7 (1): 1201. Doi:10.1038/s41598-017-01361-8

Permissions

We would like to thank the editorial team for lending their expertise to make the book truly unique. They have played a crucial role in the development of this book. Without their invaluable contributions this book wouldn't have been possible. They have made vital efforts to compile up to date information on the varied aspects of this subject to make this book a valuable addition to the collection of many professionals and students.

This book was conceptualized with the vision of imparting up-to-date and integrated information in this field. To ensure the same, a matchless editorial board was set up. Every individual on the board went through rigorous rounds of assessment to prove their worth. After which they invested a large part of their time researching and compiling the most relevant data for our readers.

The editorial board has been involved in producing this book since its inception. They have spent rigorous hours researching and exploring the diverse topics which have resulted in the successful publishing of this book. They have passed on their knowledge of decades through this book. To expedite this challenging task, the publisher supported the team at every step. A small team of assistant editors was also appointed to further simplify the editing procedure and attain best results for the readers.

Apart from the editorial board, the designing team has also invested a significant amount of their time in understanding the subject and creating the most relevant covers. They scrutinized every image to scout for the most suitable representation of the subject and create an appropriate cover for the book.

The publishing team has been an ardent support to the editorial, designing and production team. Their endless efforts to recruit the best for this project, has resulted in the accomplishment of this book. They are a veteran in the field of academics and their pool of knowledge is as vast as their experience in printing. Their expertise and guidance has proved useful at every step. Their uncompromising quality standards have made this book an exceptional effort. Their encouragement from time to time has been an inspiration for everyone.

The publisher and the editorial board hope that this book will prove to be a valuable piece of knowledge for students, practitioners and scholars across the globe.

Index